Baochang Zhang, Ce Li, Nana Lin
Machine Learning and Visual Perception

Also of Interest

Baochang Zhang, Ce Li, Nana Lin

Machine Learning and Visual Perception

—

DE GRUYTER

清華大學出版社
TSINGHUA UNIVERSITY PRESS

Authors
Baochang Zhang
School of Automation Science and Electrical Engineering
Beihang University
Beijing, China

Ce Li
School of Mechanical Electronic and Information Engineering
China University of Mining and Technology
Beijing, China

Nana Lin
Surfilter Network Technology Co., Ltd.
Haidian District
Beijing, China

ISBN 978-3-11-059553-6
e-ISBN (PDF) 978-3-11-059556-7
e-ISBN (EPUB) 978-3-11-059322-8

Library of Congress Control Number: 2019945455

Bibliographic information published by the Deutsche Nationalbibliothek
The Deutsche Nationalbibliothek lists this publication in the Deutsche Nationalbibliografie;
detailed bibliographic data are available on the Internet at http://dnb.dnb.de.

Contents

Introduction

Pattern recognition was developed in the 1920s, with the development of computer in the 1940s and the rise of artificial intelligence in the 1950s, which was rapidly developed as a discipline in the early 1960s. We need to provide a definition of patterns and pattern recognition. In the broad sense, if things observed in time and space can be distinguished based on whether they are the same or similar, they are called patterns. In the narrow sense, however, pattern is the temporal and spatial distribution information obtained through the observation of several specific characteristics. The category of patterns or the whole set of patterns in the same class is called pattern class (or simply class). Pattern recognition is the process of classifying unrecognized patterns into their corresponding pattern classes based on certain measurements or observations. After numerous years of research and development, pattern recognition technologies are now widely used in many important fields of artificial intelligence, computer engineering, robotics, neurobiology, aerospace science, weapons, and so on, such as speech recognition, speech translation, face recognition, fingerprint recognition, and biological authentication. These pattern recognition technologies are considered important and widely valued because of their role in furthering development of the national economy and defense science and technology.

Machine learning and visual perception, which are an integral part of pattern recognition, have become widely popular in the fields of computer and automation, and have become a primary area of research. With increasing numbers of industries applying machine learning to solve problems, vision – as the most important way of information acquisition – is one of the most important research fields. The authors of this book bring together their individual knowledge and long-term practice in scientific research and present this book as a professional teaching material for both undergraduate and graduate students majoring in pattern recognition who have certain mathematical bases as well as for readers interested in working on related fields of pattern recognition, including machine learning, visual perception, and so on. Machine learning algorithms, being mostly related to linear algebra and matrix theory, the authors assume that readers of this book are well conversant in these basic mathematical areas. This book introduces the main principles and methods of machine learning, as well as the latest developments. More specifically, it includes the history of machine learning, decision tree learning, the PAC model, Bayesian learning, support vector machine (SVM), AdaBoost, compression perception, subspace, deep learning, and reinforcement learning and neural network.

With numerous books focusing on machine learning and visual perception, this book has been written with the aim of introducing the basic concepts discussed in other books, as well as expound on many primary algorithms and principles used in recent years, allowing readers to not only learn this basic knowledge but also

https://doi.org/10.1515/9783110595567-001

identify the main direction it is moving in. Thus, the main works of the authors are presented as two primary aspects. On the one hand, the book presents the principles from the readers' perspective that might help them in learning, such as decision tree, Bayesian learning, SVM, compression perception, and deep learning. Unlike the tedious theories and complicated formulas in other books, this book emphasizes the practical aspects and explains a large number of implementation examples of algorithms in order to help readers understand them. On the other hand, each chapter of the book is relatively independent, which includes not only the traditional theories and methods but also novel algorithms published in recent conferences as well as popular deep learning theories, allowing students and readers alike to understand the fundamentals and new research directions in machine learning.

This book introduces the latest achievements in machine learning and reviews the research results of several authors. With the many years of studies on classifier design, face recognition, video understanding, palm pattern recognition, and railway image detection, the book would be helpful for researchers interested in the related fields.

This book is written by Baochang Zhang, Ce Li, and Nana Lin. Baochang Zhang and Ce Li are equally responsible for the content of the book, and Nana Lin contributed to the first Chinese version of this book. The authors thank Juan Liu, Lei Wang, and other graduate students who have performed much work on material collection of this book. In writing, the authors reference a large number of books related to machine learning; for more details, please refer to the key list of bibliography. Without their contribution, the book would not have been published, and the authors would like to express their sincere thanks to them.

1 Introduction of machine learning

Introduction

Machine learning is the study of how computers simulate or implement the learning behaviors of human beings to acquire new knowledge or skills and reorganize the existing knowledge structures to improve its own performance continuously (Russell, 2016) (Bishop, 2006). Machine learning, which is at the core of artificial intelligence, is the fundamental way to make computers intelligent, and its application spans all fields of artificial intelligence. It mainly uses induction and synthesis rather than deduction.

1.1 Machine learning

1.1.1 Basic concepts

Although learning ability is one of the most important characteristics of intelligent behavior, its mechanism of learning still remains unknown. Various definitions of machine learning have been propounded. H. A. Simon argues that learning is an adaptive system that becomes more effective when it completes the same or similar tasks at the next time. R. S. Michalski believes that learning is the construction or modification of the representation of what is experienced. People engaged in the development of expert systems believe that learning is the acquisition of knowledge. These views have different emphases. The first view emphasizes the external behavioral effects of learning, the second emphasizes the internal process of learning, and the third focuses on the practicality of knowledge engineering.

Machine learning, as shown in Figure 1.1, plays an important role in the field of artificial intelligence (Zhou, 2016). An intelligent system without learning ability is hardly a true intelligent system; however, most conventional intelligent systems generally lack this learning ability. For example, they cannot self-correct when they encounter an error, they do not improve their performance through experience, and they do not automatically acquire and discover the knowledge they need. Their reasoning is limited to deduction and lack of induction, so at most they can only prove that there are facts and theorems, but not new theorems, laws, and rules. With the in depth development of artificial intelligence, these limitations have become more prominent. It is in this scenario that machine learning has gradually become one of the core areas of artificial intelligence research. Machine learning has been applied to various artificial intelligence systems, such as expert system, automatic reasoning, natural language understanding, pattern recognition, computer vision, intelligent robot, and other fields (Zheng, 1998). Among them, knowledge acquisition in

https://doi.org/10.1515/9783110595567-002

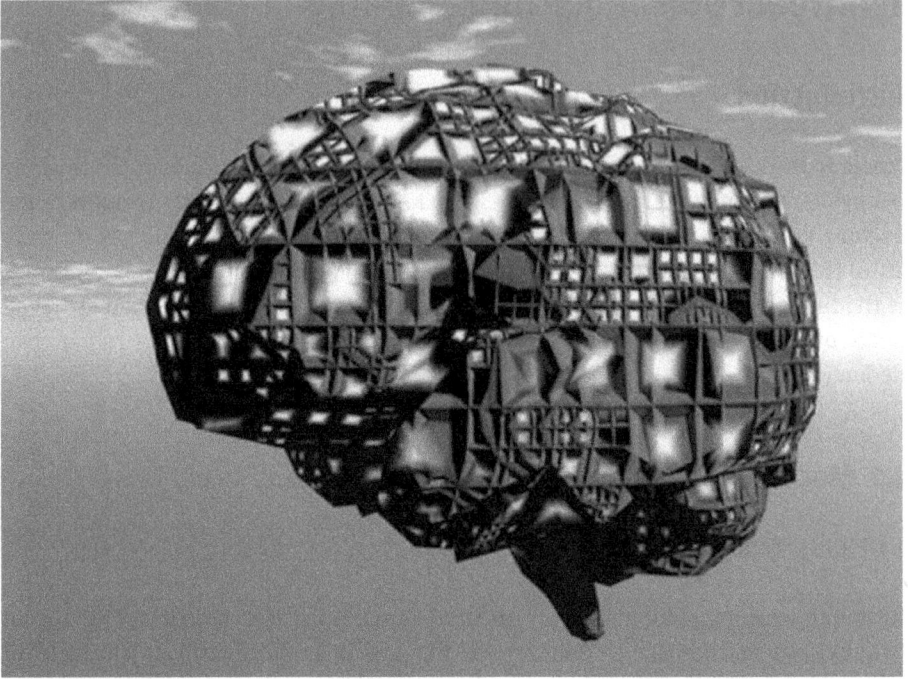

Figure 1.1: Machine learning.

expert systems is a particularly typical shortcoming, and numerous attempts have been made to overcome it using machine learning methods.

The study of machine learning is based on the understanding of human learning mechanism, such as physiology and cognitive science, in order to establish calculation models or understanding models of human learning processes, develop various learning theories and learning methods, study general learning algorithms, conduct theoretical analysis, and establish a task-oriented learning system with a specific application (Mitchell, 2013). These research goals interact with and promote each other.

Since the first machine academic seminar was held at Carnegie Mellon University in 1980, research work on machine learning has developed rapidly and has become one of the most significant topics currently.

1.1.2 Definition and significance

Learning is an important intelligent behavior possessed by human beings, and learning is a continuous process. Sociologists, logicians, and psychologists all have different opinions on learning. Currently, there is no unified definition of "machine learning," and it is difficult to provide a recognized and accurate definition. Machine learning is the study of computer algorithms that improve automatically through experience. At the same time, machine learning can also be defined as programming of computers to optimize a performance criterion using example data or past experience.

Nevertheless, to facilitate the discussion and estimate the progress of this discipline, it is necessary to define machine learning, even if this definition is incomplete and inadequate. As the name implies, machine learning is a discipline that studies how to use machines to simulate the activities of human learning. A slightly more rigorous formulation is that machine learning is a study of machines that acquire new knowledge and new skills and identify existing knowledge. Currently, the term "machine" as used herein refers to a computer or an electronic computer; however, it will likely be a neutron computer, a photonic computer, or a neural computer in the future.

Can a machine be as capable of learning as a human being? In 1959, Samuel, from the United States, designed a chess program that had the ability to learn; in other words, it could improve its chess skills step by step. Four years later, the program defeated the designer himself. Another three years later, the program defeated an American champion of eight years. This program shows the ability of machine learning and raises many thought-provoking social and philosophical issues.

Will machines surpass human beings? Researchers holding the negative answer argue that machines, which are artificial and whose activities are prescribed by the designer, will not surpass the designer in any case. This opinion about machines not having the ability to learn is partially true, but it is worth considering those machines possessing this very ability. This is because machines with learning ability constantly keep improving its applications, and in most cases the designer has no idea about the level of its ability after a while.

1.1.3 History of machine learning

Machine learning is a relatively new branch of artificial intelligence, and its development process can be divided into four periods.

The first stage was the enthusiasm period from the mid-1950s to the mid-1960s.

The second stage was the calm period of machine learning from the mid-1960s to the mid-1970s.

The third stage is from the mid-1970s to the mid-1980s, called the revival period.

The latest phase of machine learning began in 1986, which features the following aspects.

(1) Machine learning has become a new frontier discipline and a course in universities, which combines applied psychology, biology, neurophysiology, mathematics, automation, and computer science to form the theoretical basis for machine learning.

(2) Ensemble learning is on the rise, which combines various learning methods and learns from one another. In particular, connectionist learning coupled with the symbolic system is particularly noticeable because of its ability to better resolve the problems of knowledge and skill acquisition and its ability to refine continuous signal processing.

(3) An integrated viewpoint on the basic issues of machine learning and artificial intelligence is developing. The case-based approach combining analogical learning with problem-solving has become an important direction for empirical learning.

(4) The application of various learning methods is fast expanding, and some methods have been used in practical applications. Connectionist learning coupled with the symbolic system will play a role in the enterprise intelligent management and intelligent robot motion planning. Analytical learning is utilized to design a comprehensive expert system. Genetic algorithm and reinforcement learning have a bright application prospect in the field of engineering control.

(5) The academic activities related to machine learning are unprecedentedly active. Along with annually machine learning seminars, conferences are also held on computer learning theory and genetic algorithms.

1.1.4 Machine learning system

Closely related to the reasoning process, the learning process is a complex and intelligent activity. The strategies used in machine learning can be roughly divided into four types: mechanical learning, learning through learning, analogy, and learn from examples. The more the reasoning used in learning, the stronger is the system's ability.

1.1.5 Basic elements of the machine learning system

The basic elements of the machine learning system are the elements of learning system, which obtains relevant information from the environment and uses this information to modify the base knowledge for improving the execution system. The execution system carries out the tasks depending on the knowledge bases and simultaneously feeds back the obtained information to the learning system. In the specific applications, the environment, knowledge base, and the execution system determine the specific tasks, and they also entirely define the problems that will be solved in the learning system. The impacts of these three parts on the design of a learning system are described next.

The most important factor affecting the design of a learning system is the information that the environment provides to the system, or more specifically, the quality of information. The general principles for guiding the actions of the execution system are stored in the knowledge base; however, the information provided by the environment to the learning system is diverse. If there is high-quality information and small differences from general principles, the learning process will be easy to handle. If the learning system is provided with disorganized specific information that guides the information of specific actions, it will need to delete unnecessary details, summarize and popularize, form the general principle of guiding action, and release the knowledge base. However, the task of the learning part is relatively heavy and more difficult to design.

Because the information obtained by the learning system is often incomplete, the reasoning generally is not reliable, and the summarized rules may be incorrect. This needs to be tested by the feedback of the execution system. Reasonable rules can make the system more efficient, which should be preserved, whereas bad rules should be modified or removed from the database.

The knowledge base is the second factor affecting the design of a learning system. There are many forms of representation of knowledge, such as feature vector, first-order logic, production rules, semantic networks, and frameworks. These representations have their own characteristics. When choosing the means of expression, four aspects should be taken into account: strong ability to express; easy to reason; easy to modify the knowledge base; and easy to expand knowledge representation.

It is worth emphasizing the last problem that the learning system cannot acquire any information without a priori knowledge. Each learning system requires a priori knowledge to obtain information from the environment, make analyses and comparisons, make assumptions, and test and correct it. Therefore, more precisely, the learning system can expand and improve the existing knowledge.

1.1.6 Category of machine learning

1.1.6.1 Classification based on learning strategies

Learning strategy refers to the inference strategy adopted by the system in the learning process. A learning system is always composed of learning and environment. With information provided by the environment (such as books or teachers), the learning system realizes the conversion of information, records it in an understandable form, and obtains useful information. During the learning process, the less inference the students (learning part) use, the more they rely on the teacher (environment) and the heavier is the teacher's burden. The classification criteria of learning strategies are categorized according to how much and how easy it is for students to translate information (Duda, 2012), and from simple to complex, from as few as possible to the following six basic types.

(1) Rote learning

Rote learning is a method of memorization based on repetition. Here, the basic idea is that one can quickly remember the meaning of the material if one repeats it more. There are some alternative methods of rote learning, including associative learning, meaningful learning, and active learning. Rote learning is widely used to master basic knowledge. Some examples of school topics in which rote learning is often used are the acoustics of reading, multiplication tables in math, legal cases in law, anatomy in medicine, basic formulas in other science, and so on. Rote learning bypasses understanding, so it is not an efficient method to learn any complicated topic on a priority level by itself. An example of rote learning is encountered when quickly preparing for exams, also called "cramming."

Learners do not need any reasoning or other knowledge transfer to directly absorb the information provided by the environment, such as Samuel's checkers program and Newell and Simon's LT system. The main consideration of this type of learning system is indexing the stored knowledge and using it. The systematic learning method learns directly through preprogrammed and constructed programs. The learner does not do any work, or learns by directly receiving the established facts and data, and does not make any reasoning about the input information.

(2) Learning from instruction

Students obtain information from the environment (teachers or other sources of information such as textbooks, etc.), abstract it as the new knowledge, and then combine with existing knowledge. So in this learning process, students are required to have a certain degree of reasoning ability, but the environment still plays a major part. Teachers put forward and organize knowledge in some form to allow students to continually increase their knowledge. This method of learning is similar to that of human society in the school, and the task of learning is to establish a system that enables it to receive instruction and advice and to effectively store and apply the learned knowledge. Currently, many expert systems use this method to establish a knowledge base.

(3) Learning by deduction

The reasoning form used by students is deductive reasoning. Inference starts from axioms and leads to the conclusion through logical transformation. This reasoning is the process of "fidelity" transformation and specialization, which enables students to gain useful knowledge in reasoning. Learning by deduction includes macro-operation learning, knowledge editing, and chunking. The inverse process of deductive reasoning is inductive reasoning.

(4) Learning by analogy

Using the similarity of knowledge in two different domains (source, target), knowledge of the target domain can be deduced from the knowledge of analogies, masters, and domains (including similar characteristics and other properties) to enable learning. Analogical learning systems can transform an existing computer application into a new area to accomplish similar functions that were not previously designed. Analogical learning requires more reasoning than the three learning methods described previously. It generally involves retrieving the available knowledge from the source (source domain) first, then converting it into a new form, and finally using the new state (the target domain). Analogical learning plays an important role in the history of human science and technology. Many scientific discoveries have been made through analogy; for example, the famous Rutherford analogy reveals the mystery of the atomic structure using the analogy of the atomic structure (target domain) with the solar system (source domain).

(5) Explanation-based learning (EBL)
Based on the target concept, an example of this concept, the domain theory, and operational guidelines provided by the teacher, the student first constructs an explanation as to why the given example satisfies the target concept and then generalizes the explanation as a sufficient condition for the operational concept to satisfy the target concept. EBL has been widely used in the knowledge base to refine and improve system performance. Famous EBL systems include GENESIS by G. DeJong, LEXII and LEAP by T. Mitchell, and PRODIGY by S. Minton.

(6) Learning from induction
Learning from induction is a collection of examples or counterexamples of a concept provided by teachers or the environment that allows deriving a general description of the concept through inductive reasoning. This kind of learning utilizes relatively higher reasoning processes than teaching learning and deductive learning because the environment does not provide a general description of the concepts (such as axioms). To some extent, the learning process from induction inference is more than that from analogical learning, because no similar concept can be used as a "source concept." Inductive learning is the most basic and comparatively mature form of learning method, which has been widely studied and applied in the field of artificial intelligence.

1.1.6.2 Classification based on the representation of acquired knowledge
Knowledge acquired by learning systems may include behavioral rules, descriptions of physical objects, problem-solving strategies, various classifications, and other types of knowledge used for task implementation.

Following are some of the main expressions used for the knowledge gained in learning.

(1) Algebraic expression parameter

The goal of learning is to adjust the algebraic expression parameter or coefficient in a fixed function to achieve ideal performance.

(2) Decision tree

A decision tree is used to classify generic objects. Here, each internal node in the tree corresponds to an object attribute, and each side corresponds to an optional value of these attributes. The leaf node of the tree corresponds to each basic classification of the object.

(3) Formal grammar

In learning to recognize a particular language, a formal grammar of the language is formed by a series of expressions of the language.

(4) Production rules

Generative rules expressed as a pair of conditional actions have been used very widely. Learning behavior in learning systems mainly constitutes generated, generalized, specialization, or synthetic production rules.

(5) Formal logic expressions

The basic components of formal logic expressions are propositions, predicates, variables, statements that constrain the range of variables, and embedded logic expressions.

(6) Graph and network

Some systems use graph matching and graph transformation schemes to effectively compare and index knowledge.

(7) Framework and pattern

Each framework contains a set of slots that describes all aspects of things (concepts and individuals).

(8) Computer programs and other process codes

Acquiring this form of knowledge, the purpose is to obtain the ability to achieve a particular process rather than to infer the internal structure of the process.

(9) Neural networks

This is mainly used in connectionist learning. Learning acquired knowledge, and summed up as a neural network.

(10) Combination of multiple representations
Sometimes the knowledge acquired in a learning system requires the comprehensive application of several of the above-mentioned forms of knowledge.

According to the level of sophistication, knowledge representation can be divided into two broad categories: coarse-grained symbols of generalization and sub-symbols of low generalization. Similar to decision trees, formal grammar, production rules, formal logic expressions, and frameworks and patterns belong to the symbolic representation class; however, algebraic expression parameters, graphs, networks, neural networks, and so on belong to the sub-symbolic representation class.

1.1.6.3 Classification based on application area
The main areas of application include expert systems, cognitive simulation, planning and problem-solving, data mining, network information services, image recognition, fault diagnosis, natural language understanding, robotics, and gaming.

The types of tasks reflected in the implementation of machine learning show that most of the field of applied research basically concentrates on two categories: classification and problem-solving.
(1) In the classification task, the system needs to analyze the input unknown pattern (the description of the pattern) based on the known classification knowledge to determine the generic of the input pattern. The corresponding learning objective is to determine the criteria used for classification (e.g., classification rules).
(2) In the problem-solving task, the system needs to find a sequence of actions that can translate the current state into a target state for a given target state. Most of the research on machine learning in this field focuses on the lack of knowledge in order to improve the efficiency of problem-solving through learning search control knowledge, heuristic knowledge, and so on.

1.1.6.4 Comprehensive classification
All of these machine learning methods are considered together with the historical origin, knowledge representation, reasoning strategy, the similarity of result evaluation, the relative concentration of researcher exchange, the application field, and so on. They can be divided into the following six categories.

(1) Empirical inductive learning
Empirical inductive learning involves inductive learning of examples using data-intensive empirical methods (such as version space method, ID3 method, law discovery method). Examples and learning results generally use attributes, predicates, relationships, and other symbols. It is similar to inductive learning based on the classification of learning strategies, but without involving linking learning, genetic algorithm, and learning reinforcement.

(2) Analytic learning

The goal of analytical learning is not describing new concepts, but improving the performance of the system. Analytical learning includes applied interpretation learning, deductive learning, multilevel structural chunking, and macro-operations learning. Its features are summarized below:
1) the reasoning strategy is mainly deductive, rather than inductive, and
2) use of past experiences (examples) of solving problems to guide new problem-solving processes, or generating search control rules that can more effectively apply domain knowledge.

(3) Analogy learning

Analogy learning is based on learning strategy classification. Currently, the more compelling research in this type of learning involves learning by analogy with concrete examples of past experiences, known as example-based learning, or simply as sample learning.

(4) Genetic algorithm

Genetic algorithms simulate mutations in biological reproduction, exchange, and Darwinian natural selection (survival of the fittest in each ecosystem). It encodes possible solutions of the problem as a vector, called an individual. Each element of the vector is called a gene. It evaluates each individual in the population (a set of individuals) using an objective function (corresponding to the natural selection criterion) and results in a new group according to the evaluation value (fitness) of the individual selection, exchange, mutation, and other genetic operations. Genetic algorithms are suitable for very complex and difficult environments, such as large amounts of noise and irrelevant data, things that are constantly updated, problems that cannot be clearly and precisely defined, and long-running processes to determine the value of a current behavior. Similar to neural networks, the research of genetic algorithms has developed into an independent branch of artificial intelligence, pioneered by J. H. Holland.

(5) Connectionist learning

A typical connection model is implemented as an artificial neural network consisting of a weighted connection between the unary elements called neurons.

(6) Reinforcement learning

Reinforcement learning is characterized by tentative interaction with the environment to determine and optimize the choice of actions in order to achieve sequence decision tasks. In this task, the learning mechanism interacts with the environment by selecting and executing actions that lead to changes in the state of the system and possibly to some kind of fortified signal (immediate reward), which is a measure of the behavior of the system's quantified reward and punishment. The learning goal is to find a suitable action selection strategy to help to choose the action in any given state, so as to obtain a certain optimal result for the generated action sequence (e.g., the accumulated immediate return).

Comprehensive classification, such as inductive learning, genetic algorithm, connected learning, and reinforcement learning, falls under inductive learning, where the induction of learning experience uses symbolic representation, and genetic algorithm, connections-based learning, and reinforcement learning use sub-symbolic representation; however, analysis of learning falls under deduction learning.

In fact, the analogy strategy can be considered as a combination of induction and deductive strategy. Therefore, the most basic learning strategies are only induction and deduction.

Learning by inductive strategy involves inductive input, where the knowledge acquired is far greater than that in the original system knowledge base. The result obtained in such a way changes the knowledge deduction closure of the system, and hence this kind of learning can be called knowledge-based learning. The knowledge acquired using deductive strategies can improve the efficiency of the system, but still root in the original system of knowledge base, that is, the acquired knowledge cannot change the system of deduction closure; hence, this type of learning is also known as symbol-level learning.

1.1.7 Current research field

At present, research in the field of machine learning mainly focuses on the following three aspects:
(1) task-oriented research: study, analyze, and improve learning systems having executive performance of a set of scheduled tasks;
(2) cognitive model: study the human learning process and conduct computer simulation; and
(3) theoretical analysis: theoretically explore various possible learning methods, independent of the application of the algorithm.

Machine learning is another important field of artificial intelligence applications following the expert system and one of the core research topics in artificial intelligence and neural computing. Existing computer and artificial intelligence systems have little or, at best, only a very limited learning ability, thus failing to meet the new requirements of technology and production. The discussion and progress of machine learning will further promote the development of artificial intelligence as well as science and technology on a whole.

1.2 Statistical pattern recognition

The problem of statistical pattern recognition can be considered a special case of a broader problem, that is, data-based machine learning problems (V., 2004, 1998). Data-based machine learning is a very important aspect of modern intelligent technology. It mainly studies the derivation of laws that cannot be obtained through the principle analysis from some observation data (samples), uses these laws to analyze the objective objects, and predicts the future data or unobservable data. In the real world, there are a vast number of things that people cannot know about accurately but can only observe. Therefore, this kind of machine learning has very important applications in various fields ranging from modern science and technology to society and economy. If we focus on the classification of the input object, the machine learning problem can be called pattern recognition. This chapter will discuss pattern recognition under the larger framework of data-based machine learning and will be referred to as machine learning for short.

Statistics are the most basic (and only) means of analysis when people are dealing with data and lack theoretical models and are the basis for the various approaches presented in this chapter. Traditionally, statistics have studied the asymptotic theory, which is the ultimate characteristic when the number of samples tends to infinity. The bounds of statistical consistency, meta-deviance, and estimated variance in statistics, as well as the previously discussed many conclusions of classification error rates, belong to this asymptotic property. However, in practical applications, such preconditions are often not satisfied. This is especially true when the problem is in a high-dimensional space. This is actually the case with existing machine learning theories and methods, including pattern recognition and neural networks.

Vladimir N. Vapnik et al. studied machine learning problems in the context of finite samples as early as in the 1960s, along with earlier studies in this field. Since these studies were not yet perfect at that time, they tended to be more conservative and less mathematical in solving the problem of pattern recognition. Until the 1990s, no better method was suggested to put their theory into practice. Coupled with the rapid development of other learning methods at that time, these studies have not received sufficient attention. Until the mid-1990s, the study of the machine learning theory with limited samples gradually improved, leading to a relatively perfect theoretical system – the statistical learning theory (SLT). At the same time, research on the more emerging machine learning methods, such as neural network, encountered some major difficulties, such as determining the network structure, over-learning and under-learning, local minimum problem, and so on. Under such circumstances, attention is gradually being paid to the SLT, which attempts to study machine learning problems more fundamentally.

From 1992 to 1995, based on SLT, a new pattern recognition method – the support vector machine (SVM) – was developed, which proved to have several unique

advantages in solving small-sample, nonlinear, and high-dimensional pattern recognition problems and can be extended to other machine learning problems, such
as function fits. Because many problems still exist in SLT and SVM methods, further
research is needed. However, currently, scholars are of the opinion that these methods are becoming new research hot spots in machine learning after pattern recognition and neural network research, and they will go a long way in promoting the
development of the theory and technology of machine learning.

1.2.1 Problem representation

The basic framework of machine learning is shown in Figure 1.2. Here, the system S
denotes that providing a certain input$x$ aims at achieving an output y, where LM is
the learning machine whose output is $\hat{y}$. According to a given training sample, the
learning machine is seeking to estimate the dependency between the input and the
output, making it possible to predict the unknown output as accurately as possible.

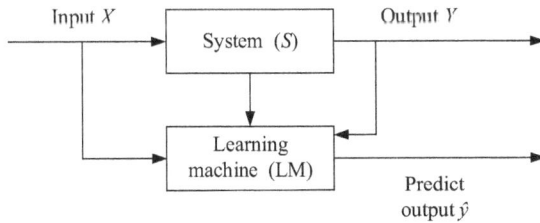

Figure 1.2: Basic framework of machine learning (ZhaoQi, 2012).

Machine learning problems can be formally expressed as follows. There is a certain
unknown relationship between the variable y and the input x, that is, there exists
an unknown joint probability $F(x,y)$ (the deterministic relationship between x and y
can be as a special case); machine learning is based on n independent observation
samples with the same distribution.

$$(x_1, y_1), (x_2, y_2), ..., (x_n, y_n) \tag{1.1}$$

Find the optimal function $f(x, \omega_0)$ in a set of functions $\{f(x, \omega)\}$, and minimize the
expected risk of the forecast as

$$R(\omega) = \int L(y, f(x, \omega)) dF(x, y) \tag{1.2}$$

Where $\{f(x, \omega)\}$ is a set of predictive functions and $\omega \in \Omega$ is a generalized parameter of the function so that $\{f(x, \omega)\}$ can represent any set of functions; $L(y, f(x, \omega))$
is the loss of forecasting y in $\{f(x, \omega)\}$. The different types of learning problems

have different forms of loss functions; predictive functions are often referred to as learning functions, learning models, or learning machines.

There are three basic types of machine learning problems: pattern recognition, function approximation, and probability density estimation.

For pattern recognition problems (where only supervised pattern recognition problems are discussed), the system output is the class label. In both cases, $y = \{0, 1\}$ or $\{-1, 1\}$ is a binary function. The predictive function is then called the indicator function, which is the discriminant function mentioned later in this book. The basic definition of the loss function in a pattern recognition problem can be

$$L(y, f(x, \omega)) = \begin{cases} 0 & if \quad y = f(x, \omega) \\ 1 & if \quad y \neq f(x, \omega) \end{cases} \tag{1.3}$$

With the definition of loss function, minimizing the expected risk, the average error rate discussed in Chapter 2, is solved by a pattern recognition method – Bayesian decision-making. Other loss functions are defined to obtain other decision-making methods.

Similarly, in the fitting problem, if y is a continuous variable (here assumed to be a single-valued function), which is a function that depends on x, then the loss function can be defined as

$$L(y, f(x, \omega)) = (y - f(x, \omega))^2 \tag{1.4}$$

In fact, as long as the output of a function is transformed into a binary function by a threshold, the fitting problem becomes a problem of pattern recognition.

For the probability density estimation problem, the purpose of learning is to determine the probability distribution of x according to the training samples, and the estimated density function is $p = (x, \omega)$. The loss function can be defined as

$$L(p(x, \omega)) = -\log p(x, \omega) \tag{1.5}$$

1.2.2 Experience risk minimization

Obviously, to minimize the expected risk defined in eq. (1.2), we must rely on the information about joint probability $F(x, y)$. It is necessary to know the prior probability and the conditional probability density in the pattern recognition problem. However, in the actual machine learning problem, only the information of the known sample eq. (1.1) can be used, so the risk and the meta-method are expected to be calculated and minimized directly.

According to the idea of the theorem of large numbers in probability theory, the mean is substituted for the expectation in eq. (1.2), thus defining to approximate the expected risk as defined by eq. (1.2). Since $R_{e\,mp}(\omega)$ is defined using known training samples (i.e., empirical data), it is called empirical risk. Substituting the minimum value of the empirical risk for the parameter ω for the minimum value of the expected risk $R_{e\,mp}(\omega)$ is called the empirical risk minimization (ERM) principle. The various data-based classifier design methods introduced earlier are actually proposed under the principle of minimizing empirical risk.

$$R_{e\,mp}(\omega) = \frac{1}{n}\sum_{i=1}^{n} L(y_i, f(x_i, \omega)) \tag{1.6}$$

In the function fitting problem, the traditional least squares fitting method is obtained by substituting the loss function defined in eq. (1.4) into eq. (1.6) and minimizing the empirical risk. However, in the probability density, the ERM method using the loss function of eq. (1.5) is the maximum likelihood method.

Carefully studying the principle of minimizing empirical risk and minimizing expected risk in machine learning problems, we find that there is no reliable theoretical basis for minimizing both expected risk and empirical risk, which is only an intuitively reasonable approach.

First, both $R_{e\,mp}(\omega)$ and $R(\omega)$ are functions of ω. The large number theorem in the probability theory shows (under certain conditions) that $R_{e\,mp}(\omega)$ approaches the value of $R(\omega)$ when the sample tends to be infinitely long, which is no guarantee that the minimum ω^* of $R_{e\,mp}(\omega)$ and the minimum ω'^* of $R(\omega)$ are the same point, and there is no guarantee that $R_{emp}(\omega^*)$ can approach $R(\omega'^*)$.

Second, even if there are ways to ensure that these conditions are guaranteed when the number of samples is infinite, it cannot be assumed that the ERM approach from these assumptions yields good results with a limited sample size.

Despite these unknown issues, the minimization of empirical risk as the basic idea for solving machine learning problems such as pattern recognition still dominates almost all studies in this area, and for many years, studies have mostly focused their attention on better determining the least empirical risk. In contrast, the theory of statistical learning is an in-depth study of the basic issues including the prerequisites for the expected risk minimization using the principle of solving ERM. What is the performance of the ERM approach if these are not true, and if more reasonable rules can be found?

1.2.3 Complexity and generalization

In early neural network studies, researchers always focused on how to make R smaller but soon discovered that blind pursuit of small training errors does not always achieve good predictive results. The ability to make the machine learn how to predict the future output correctly is called generalization. In some cases, training error is so small that it leads to the decline of generalization; this is called the over-fitting problem, which almost all neural network researchers have encountered at some point of time. Theoretically, the same problem exists in pattern recognition. However, because the commonly used classifier models are relatively simple (such as linear classifiers), over-learning problems are not as prominent as in neural networks.

There are two reasons leading to the phenomenon of over-fitting: first, the learning sample is not enough, and second, the learning machine design is unreasonable. These two issues are interrelated. Just imagine a very simple example, assuming that there is a set of training samples (x, y), x distributed in the real range, and y values between $[0, 1]$. Then, regardless of the function model these samples are based on, simply $f(x, a) = \sin(ax)$ is used to fit these samples, where a is the undetermined parameter and always finds an a such that the training error is zero; however, the "optimal function" cannot correctly represent the original function model. This happens because trying to fit a limited sample with a complicated model results in the loss of the generalization. For neural network, if the learning ability of the network is too strong for a limited number of training samples, at this time the risk of experience can quickly converge to a small or even zero, but we simply cannot guarantee that it will be able to get a good prediction of the new sample in the future. This is the contradiction between the complexity and the generalization of a learning machine for a finite set of samples.

In many cases, even if the samples in the known problem come from a relatively complex model, the learning effect on the sample with the complicated prediction function is usually not as good as that of the relatively simple prediction function due to limited training samples. This is even more so when there is noise. For example, ten samples are generated using the quadratic model $y = x^2$ under noisy conditions and fitted by a linear function and a quadratic function, respectively, based on the principle of minimizing empirical risk. Although the real model is a quadratic polynomial, the result of a one-time polynomial prediction is closer to the real model due to the limited number of samples and the noise impact. The same experiment was carried out 100 times, and 71% of the experimental results were found to be better than a quadratic fitting. The same phenomenon can easily be seen in the pattern recognition problem.

From these discussions, it is possible to draw the following basic conclusion: in the case of finite samples, the least empirical risk does not necessarily mean that the expected risk is minimal; the complexity of the learning machine is relevant not only to the system under study but also to the limited learning samples.

The contradiction between learning accuracy and generalizability appears to be irreconcilable in the case of finite samples. The use of complex learning machines tends to make learning errors smaller but tends to lose generality. Therefore, many original methods have been improved by studying many remedies, such as penalizing the complexity of learning functions in training errors or by selecting models for controlling complexity through cross-validation. However, these methods are more experience-based and hence lack a sound theoretical basis. In the study of neural networks, specific problems can be studied by reasonably designing a network structure and a learning algorithm to achieve both learning precision and generalization. However, no guiding theory exists on how to do it. In pattern recognition, people tend to adopt simpler classifier models such as linear or piecewise linear.

1.3 Core theory of statistical learning

SLT is considered the best theory for statistical estimation and prediction learning of small samples. It systematically studies the conditions for establishing the principle of minimization of empirical risk, the relationship between empirical risk and expected risk under a limited sample, and how to use these theories to find new learning principles and methods. The main contents include the following four aspects:
(1) the conditions of statistical learning consistency under the principle of ERM;
(2) conclusions about the generality of statistical learning methods under these conditions;
(3) the principle of reasoning based on few shot samples; and
(4) practical ways to implement these new principles (algorithms).

1.3.1 Consistency condition of the learning process

The conclusion about learning consistency is the basis of SLT as well as its basic connection with traditional asymptotic statistics. The consistency of the learning process indicates that when the number of training samples tends to infinity, the optimal value of empirical risk can converge to the optimal value of real risk. Only by satisfying the condition of consistency can we ensure that the optimal method obtained under the principle of ERM approaches the optimal result that minimizes the expected risk when the sample is infinite.

1.3.2 Generalization bounds

The previous discussion leads us to a series of conditions for learning the convergence and convergence speed of machines. Although theoretically significant, in practice, they are generally not directly applicable. Here, we will discuss the important conclusions of SLT regarding the relationship between empirical risk and real risk – called generalized ones – which are important bases for analyzing machine performance and developing new learning algorithms.

Because the function set has a finite Vapnik–Chervonenkis (VC) dimension, which is a necessary and sufficient condition for consistent convergence of the learning process, only the limited functions of the VC dimension are discussed here unless otherwise specified.

According to the conclusion of the SLT about the generalization of function sets, if the value of loss function $Q(x, w) = L(y, f(x, w))$ is 0 or 1, then the following theorem can be arrived at.

Theorem 1.1: For the two types of classification problems defined previously, all the functions in the set of indicated functions (of course, functions that minimize the empirical risk), the empirical risk, and the actual risk satisfy at least the probabilities $1 - \eta$ for the following relationships:

$$R(\omega) \leq R_{emp}(\omega) \frac{1}{n} + \frac{1}{2}\sqrt{\varepsilon} \tag{1.7}$$

Here, when the function set contains an infinite number of elements (i.e., the parameter ω has an infinite number of possible values)

$$\varepsilon = \varepsilon\left(\frac{n}{h}, \frac{-In\eta}{n}\right) = a_1 \frac{h\left(In\frac{a_2 n}{h} + 1\right) - In(\eta/4)}{n} \tag{1.8}$$

And when the function set contains a finite number of (N) elements

$$\varepsilon = 2\frac{InN - In\eta}{n} \tag{1.9}$$

Where h is the VC dimension of the function set. In general, there are infinitely many possible classifiers, so we use eq. (1.8), where a_1 and a_2 are two constants that satisfy $0 < a_1 \leq a_4, 0 < a_2 \leq 2$. In the worst case, the relationship $a_1 = 4, a_2 = 2$ can be further simplified as

$$R(\omega) \leq R_{emp}(\omega) + \sqrt{\frac{h(In(2n/h) + 1) - In(\eta/4)}{n}} \tag{1.10}$$

If the loss function $Q(x, w)$ is a general bounded non-negative real function, that is, $0 \leq Q(x, w) \leq B$, then the following theorem can be arrived at.

Theorem 1.2: For all functions in a function set (including functions that minimize empirical risk), the following relationships are at least in probability $1 - \eta$:

$$R(\omega) \leq R_{emp}(\omega) + \frac{B\varepsilon}{2}\left(1 + \sqrt{1 + \frac{4R_{emp}(\omega)}{B\varepsilon}}\right) \tag{1.11}$$

Here, ε is still defined by eq. (1.9).

When the loss function is unbounded, there are corresponding conclusions, which are not discussed here.

Theorem 1.1 and Theorem 1.2 tell us that the actual risk of a machine learning under the principle of minimizing empirical risk is composed of the following two parts:

$$R(\omega) \leq R_{emp}(\omega) + \varphi \qquad (1.12)$$

where the first part is the empirical risk of training samples and the second part is called the confidence range, also called VC confidence .

Equations (1.10) and (1.11) show that the confidence is affected not only by the confidence level $1 - \eta$ but also by the VC dimension of the function set and the number of training samples; additionally, as it increases, the confidence monotonically decreases. To emphasize this, eq. (1.12) can be reformulated as

$$R(\omega) \leq R_{emp}(\omega) + \varphi \left(\frac{n}{h} \right) \qquad (1.13)$$

Since Theorem 1.1 and Theorem 1.2 give the upper bounds on the differences between empirical and real risks, they reflect the generalization of learning machines based on the principle of ERM and are therefore called generalizations of the world.

Through further analysis, we can find that when the n/h is small (e.g., less than 20, then we say that the number of samples is small), the confidence range ψ is large, and the empirical risk approximates the real risk with greater error. If the sample number is large and the n/h is larger, the confidence range will be small, and the optimal solution to minimize the empirical risk approaches the actual optimal solution.

On the other hand, for a specific problem, when the sample number n is fixed, the higher the VC dimension of the learning machine (classifier) (i.e., the higher the complexity), the greater is the confidence range, resulting in a larger difference between true risk and empirical risks. Therefore, when designing a classifier, not only the empirical risk needs to be minimized but also the VC dimension should be made as small as possible, thereby reducing the confidence range to minimize the expected risk. This is why, in general, selecting an overly complex classifier or neural network often does not provide good results.

We take an example of fitting an arbitrary point with a *sin* function. The VC dimension of the *sin* function is infinite, and the empirical risk reaches zero. However, the actual risk is large and does not have any generalization. Also in the examples of Figure 1.3, although it is known that the samples are generated by quadratic functions, since fewer training samples are used, fitting with a smaller VC dimension function (making h/n smaller) yields better effect. Similarly, the phenomenon of over-fitting occurs in some methods such as neural networks because, in the case of a limited sample, if the network or algorithm is not properly designed, the empirical risk will be smaller while the confidence range will be large, resulting in a decline in generalization.

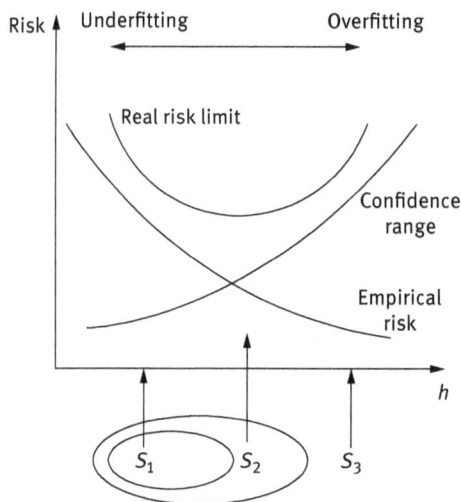

Figure 1.3: Risk minimization diagram (Zhang, 2010). Function Set Subset: $S_1 \subset S_2 \subset S_3$ VC dimension: $h_1 \leq h_2 \leq h_3$.

A point to be noted is that, similar to the key theorem of learning theory, the boundary of generalization is also the worst-case one. The given boundary is very relaxed in many cases, especially when the VC dimension is high. Moreover, it is often only effective when comparing the same type of learning function, which can help select the best function from the function set, but not necessarily set up in different function sets. In fact, searching for better parameters that can help enhance a machine's learning ability to lead to a better result is one of the important research directions of future learning theory.

Here, we discuss in particular the k nearest neighbor algorithm. Because the algorithm determines that for any training set, it is always possible to find an algorithm that correctly classifies any of these samples (e.g., in the simplest case, one nearest neighbor), the VC dimension of the k nearest neighbor classifier is infinite. But why is this algorithm usually able to achieve better results? Is it contradictory to the conclusion reached here? In fact, this is not the case, but because the neighbor's algorithm itself does not use the principle of minimizing empirical risk, the conclusions discussed here cannot be applied.

1.3.3 Structural risk minimization

The previous discussion showed that the principle of minimizing the empirical risk commonly used in traditional machine learning methods is not reasonable when the number of samples is limited because of the need to minimize both the empirical risk and the confidence range. In fact, in the traditional method, the process of selecting learning models and algorithms involves optimizing the confidence range.

If the selected model is more suitable for the existing training samples (equivalent to h/n), better results can be obtained. For example, in neural networks, different network structures (corresponding to different VC dimensions) need to be selected depending on the specific situation of the problem and the sample, thus minimizing the empirical risk.

In pattern recognition, a form of classifier (such as a linear classifier) is chosen to determine the VC dimension of the learning machine. In eq. (1.13), we first determine Φ by choosing a model, and then Φ is fixed; the minimum risk is then obtained by minimizing the empirical risk. Because of the lack of awareness of Φ, this choice is often based on prior knowledge and experience, resulting in overdependence on user "tricks" such as neural networks. Although the problem of pattern recognition is not linear in many cases, when the number of samples is limited, the linear classifier can often become a non-anchor. This is because the linear classifier has a lower VC dimension, which is best for obtaining a small range of confidence.

With the theoretical basis of eq. (1.13), another strategy can be used to solve this problem by first decomposing the function set $S = \{f(x, \omega), \omega \in \Omega\}$ into a sequence of function subsets (called substructures):

$$S_1 \subset S_2 \subset \cdots \subset S_k \subset \cdots \subset S \qquad (1.14)$$

This allows each subset to be arranged in accordance with the size of Φ, which in turn is arranged depending on the size of the VC dimension, that is

$$h_1 \leq h_2 \leq \cdots \leq h_k \leq \cdots \qquad (1.15)$$

A reasonable function subset should satisfy two basic conditions: first, the VC dimension of each subset is finite and satisfies the relationship of eq. (1.15) and, second, the loss function corresponding to the function in each subset is a bounded non-negative function, or satisfies the following relationship for a certain parameter pair (p, τ_k):

$$\sup_{\omega \in \Omega} \frac{[\int Q^p(z, \omega) dF(z)]^{\frac{1}{p}}}{\int Q(z, \omega) dF(z)} \leq \tau_k, p > 2 \qquad (1.16)$$

With the principle of structural risk minimization, the design process of a classifier includes the following two tasks:
(1) choose a suitable subset of functions so as to have the best classification ability for the problem; and
(2) choose a discriminant function from this subset to minimize empirical risk.

The first step corresponds to the choice of the model, whereas the second step corresponds to the estimation of the parameters after the formation of the function has been determined. Unlike the traditional approach, here the model is chosen through an estimate of its generalized universe.

The principle of structural risk minimization provides us with a more scientific learning machine design principle than ERM. However, since its ultimate objective is the compromise between the two sums of eq. (1.13), it is actually not easy to implement this principle. Finding a way to partition subsets without having to calculate them one by one can allow one to know the smallest possible empirical risk in each subset (e.g., all subsets can correctly classify training sample sets with a minimum empirical risk of 0). The above-mentioned two-step task can be carried out separately, by first selecting the subset that minimizes the confidence and then choosing the best one among them.

We discussed here how to construct the function subset. Unfortunately, there is currently no general theory on constructing a subset of predictive functions. SVM can prove better in minimizing the risk in a sequential way; for other examples on building subsets, readers can refer to the literature.

Summary

This chapter first introduces the basic concepts of machine learning, including the definition and research significance of machine learning, the history of machine learning, the main strategy of machine learning, the basic structure of a machine learning system, and classification and current research areas of machine learning.

Then, the basic theories of statistical pattern recognition are introduced, which are the expression of machine learning problems, minimization of experience risk, complexity, and generalization, thus laying the foundation for the further introduction of these theories later in this book.

Next, the core content of SLT is introduced, including the conditions for the consistency of the learning process, the generalization of the community, and structural risk minimization.

Just go through this chapter and get some inspiration. Some specific methods of machine learning and SLT have been described in the following chapters.

2 PAC model

Introduction

In the computational learning theory, probably approximately correct learning (PAC learning) is a framework for mathematical analysis of machine learning. It was proposed in 1984 by Leslie Valiant (Valiant, A theory of the learnable, 1984) (Haussler, 1990) (Haussler, 1993). In this framework, the learner receives samples and must select a generalization function (called the hypothesis) from a certain class of possible functions. The objective is that, with high probability (the "probably" part), the selected function will have low generalization error (the "approximately correct" part). The learner must be able to learn the concept given any arbitrary approximation ratio, probability of success, or distribution of samples. The model was later extended to treat noise (misclassified samples). An important innovation of the PAC framework is the introduction of computational complexity theory concepts to machine learning. In particular, the learner is expected to find efficient functions (time and space requirements bounded to a polynomial of the example size), and the learner itself must implement an efficient procedure (requiring an example count bounded to a polynomial of the concept size, modified by the approximation and likelihood bounds). This chapter introduces the basic PAC model and further discusses the sample complexity problem in both finite and infinite space. The discussion in this article will limit the concept of learning Boolean values, and the training data is noiseless (many of the conclusions extend to the more general case) (Valiant, Probably Approximately Correct: NatureÕs Algorithms for Learning and Prospering in a Complex World, 2013).

2.1 Basic model

2.1.1 Introduction of PAC

The main topics covered by PAC include the PAC learnability or probably learning theory, sample complexity for finite hypothesis spaces, the sample complexity for infinite hypothesis spaces mistake bound model, and learning algorithms for specific problems (Natarajan, 1991) (Kearns, 1994). Although it can also be extended to describe problems such as regression and multi-category, the original PAC model was proposed for the bi-class problem. Likewise, we have an input space X, also known as the instance space. A concept c on X is a subset of X, or simply C as a function from X to $\{0,1\}$. Obviously, c can be characterized by all those points whose function values are equal to 1; those points make up a subset of X corresponding to a one-to-one mapping "function."

https://doi.org/10.1515/9783110595567-003

2.1.2 Basic concepts

Instance space refers to all the instances that the learner can see, and x_n indicates every instance set of learning problems of size n. Each $x \in X$ is an instance; $X = U_n \geq 1, X_n$ is the instance space. Concept space refers to the set of all concepts where the target belongs to. The goal of the learner is to create a hypothesis h that can accurately classify each instance; for each $n \geq 1$, define each $C_n \subseteq 2^{x_n}$, as a series of concepts on X_n, $C = U_n \geq 1$, where C_n is the conceptual space on X, also known as concept classes. The hypothesis space algorithm can output all the assumptions h, leading to the set H. For each objective concept $c \in C_n$, and instance $x \in X_n$, $c(x)$ as the classification value on instance x, that is, $c(x) = 1$ if and only if $x \in C_0 C_n$; any one hypothesis h refers to a rule, that is, for a given $x \in X_n$, the algorithm outputs a prediction for $c(x)$ in the polynomial time. Version space is the set of all hypotheses that correctly classified training example D, $VS = \{h \in H | \forall < x, c(X) > \in D(h(X) = c(X))\}$. The significance of version space is that each consensus learner outputs a hypothesis that belongs to the version. Sample complexity refers to the minimum number of training samples required when the learner converges to a successful hypothesis. Computational complexity indicates the computational cost required for converging to a hypothesis successfully. In a particular hypothetical space, given the samples, if a hypothesis h is consistent for any concept and its computational complexity keeps polynomial, the algorithm is called a consistent algorithm.

2.1.3 Problematic

Let $X = \{0,1\}^n$ be a set called the instance space, and the encodings of all the samples, the concept class, and hypothesis space are the subsets of $\{0,1\}^n$. Consider an accuracy $\varepsilon(0 < \varepsilon < 1/2)$, a confidence degree $\delta(0 < \delta < 1)$, all distributions D in the instance space, and all the objective functions t in the target space. If the learner L only needs polynomial $P(n, 1/\varepsilon, 1/\delta)$ within the time of the polynomial $P(n, 1/\varepsilon, 1/\delta)$, there is a probability that at least $(1 - \delta)$ will eventually be output assuming that $h \in H$ such that the random sample is misclassified $error_D(h, t) = p_r[\{x \in X : h(x) \neq t(x)\}] \leq \varepsilon$. Learner L is called PAC-learnable, which is a basic framework to consider sample complexity and computational complexity. We can also say that the learner L is a PAC learning algorithm for the concept class c.

Suppose h is a binary function based on X. We try to approximate c by h and choose a probability distribution μ on X. According to the definition of error (risk), we have $\varepsilon(h) = \mu(h(X) \neq c(X))$, and denote it with an easy and intuitive concept in the set theory, called symmetric difference, as follows: $\varepsilon(h) \neq \mu(h \Delta c)$. As shown in Figure 2.1, the error is intuitively described as the area of symmetric difference between two sets (shaded area).

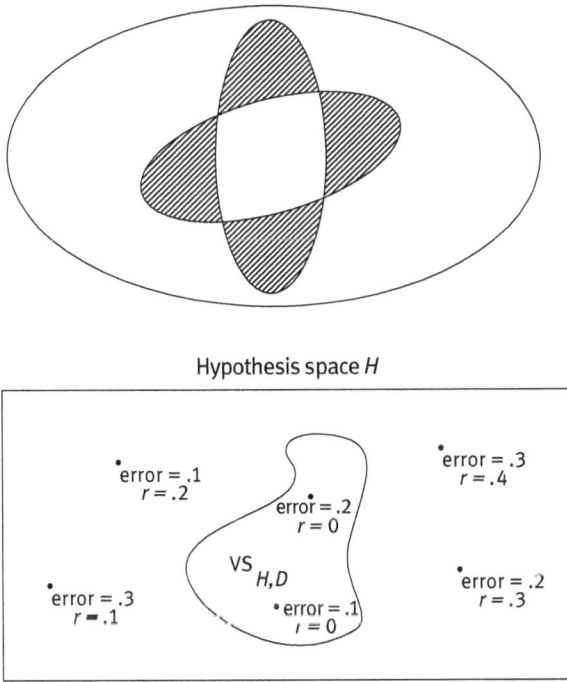

Hypothesis space H

Figure 2.1: Error risk diagram.

A concept C is a subset $c \subset X$ corresponding to the above-mentioned function space F. Similarly, the learning problem is actually a problem to fit an approximation $h \in C$ given a target concept $c \in C$. The PAC model is distribution independent because the instance distribution is unknown for the learner. This definition requires the learner to output a hypothesis with a small error less than $\varepsilon(\varepsilon$ is arbitrarily small) instead of a zero error; it also limits the randomly sampling failure rate of the learner to less than $\delta(\delta$ is arbitrarily small) instead of zero, in order to learn an approximation of correct hypothesis.

2.2 Sample complexity in the PAC model

2.2.1 Sample complexity in finite space

We start with a good class of learner – a consistent learner, defined as the one that outputs a hypothesis that perfectly fits the training data set, whenever possible. We recall that version space $VS_{H,D}$ is defined to be the set of all hypotheses $h \in H$ that correctly classify all training examples in D. Every consistent learner outputs a hypothesis belonging to version space. $VS_{H,D}$ is said to be ε-exhausted with respect to

c and D if for any h in $VS_{H,D}$, $error_D(h) < \varepsilon$. If hypo space H is finite, and D is a sequence of m independent randomly drawn examples of some target concept c, then for any $0 \le \varepsilon \le 1$, the probability that $VS_{H,D}$ is not ε-exhausted with respect to c is no more than $|H|e^{-\varepsilon m}$. The basic idea behind the proof: since H is finite, we can enumerate hypotheses in $VS_{H,D}$ by h_1, h_2, ... h_k. $VS_{H,D}$ is not ε-exhausted if at least one of these h_i satisfies $error_D(h) \ge \varepsilon$; however, such h_i perfectly fits the m number of training examples. The details will be elaborated next.

Suppose for a learner L, the hypothesis space and concept space are the same, that is, $H = C$, because the assumption space is the combination of n Boolean words, and each word has three possibilities: the variable is included in the hypothesis as a word, the negative variable is included in the hypothesis as a word, and the variable is excluded in the hypothesis. Thus, the size of the space is $|H| = 3^n$. The following algorithm can be designed.

Initialization hypothesis h is a combination of $2n$ characters, that is: $h = x_1\bar{x}_1 x_2 \bar{x}_2 \cdots x_n \bar{x}_n$.

Generate $m = 1/2(n \ln 3 + \ln 1/\delta)$ samples from the sample generator, for each positive case, delete x_i from h if $x_i = 0$; delete $\bar{x}_i$ from h if $x_i = 1$.

Output retained hypothesis h.

To analyze the algorithm, three things need to be considered: whether the number of samples are needed for the polynomial; Whether the algorithm runs in polynomial time, that is, both are $P(n, 1/\varepsilon, 1/\delta)$; and Whether the output hypothesis meets the criteria of the PAC model or not, that is $P_r[error_D(h) \le \varepsilon] \ge (1-\delta)$. For this algorithm, because the number of samples is known, it is obviously polynomial; since the time for running each sample is a constant and the number of samples is polynomial, the running time of the algorithm is also polynomial; thus, we need to just look at whether it meets the criteria of the PAC model. If we assume that h' satisfies $error_D(h') > \varepsilon$, we call it the $\varepsilon - bad$ assumption; otherwise, we call it the $\varepsilon - exhausted$ hypothesis. If the final output hypothesis is not $\varepsilon - bad$ assumption, then this hypothesis will satisfy the criteria of the PAC model.

According to the definition of $\varepsilon - bad$ assumptions: Pr[there exists a sample cosistent with the $\varepsilon - bad$ assumption] $\le (1-\varepsilon)$ is for each sample independently, and Pr [there exists m samples cosistent with the $\varepsilon - bad$ assumption] $\le (1-\varepsilon)^m$ for m samples. By incorporating the largest number of hypotheses $|H|$, we obtain Pr[there exists m samples cosistent with the $\varepsilon - bad$ assumption] $\le |H|(1-\varepsilon)^m$. Because Pr[H is a hypotheses of $\varepsilon - $bad] $\le \delta$, we have

$$|H|(1-\varepsilon)^m \le \delta$$

The solution is the following equation:

$$m \ge \frac{\ln|H| + \ln 1/\delta}{-\ln(1-\varepsilon)} \tag{2.1}$$

According to the Taylor expansion, $e^x = 1 + x + \frac{x^2}{2} + \cdots > 1 + x$; by incorporating $x = -\varepsilon$ into the Taylor expansion, we have $\varepsilon < -\ln(1-\varepsilon)$. Then, by substituting it into eq. (2.1), we have

$$m > \frac{1}{\varepsilon}\left(\ln|H| + \ln\frac{1}{\delta}\right) \tag{2.2}$$

This equation indicates a general theoretical bound of number of training examples, which is sufficient for any consistent learner to successfully learn any objective function in H at the expectation values of ε and δ. The equation indicates that the number of training examples m is sufficient to ensure that any consistent hypothesis is likely (the possibility is $(1-\delta)$) to be approximate (error rate ε). m increases linearly with $1/\varepsilon$, with the logarithm of the scale of $1/\delta$ and the assumed space.

For this case, $|H| = 3^n$. Substituting it into eq. 2.2, we obtain that $\Pr[error_D(h) > \varepsilon] \leq \delta$ holds when the number of samples $m > \frac{1}{\varepsilon}(n\ln 3 + \ln\frac{1}{\delta})$. At the same time, it also proves that the combination of the Boolean variable is PAC learnable, which is different from PAC unlearnable, such as k-term-C NF or k-term-DNF. Since eq. (2.2) uses $|H|$ to describe the sample complexity, it has the disadvantage of a loose bound; moreover, eq. (2.2) cannot be used at all for infinite hypothetical space. Therefore, it is necessary to introduce another measure – the VC dimension.

2.2.2 Sample complexity in infinite space

The VC dimension instead of $|H|$ can be used to obtain the sample complexity bound. The sample complexity based on the VC dimension is more compact than that of $|H|$, and the sample complexity of infinite hypothetical space can also be described.

In the Vapnik–Chervonenkis (VC) theory, the VC dimension is a measure of the capacity of functional space, such as complexity, richness, expressive power, or flexibility. It was originally defined by Vladimir Vapnik and Alexey Chervonenkis as the cardinality of the largest set of points that the algorithm can break up, and it can be studied by statistical classification algorithms.

Formally, the capacity of a classification model is determined by how complicated it can be. For instance, considering the threshold of a high-degree polynomial, if the value of the polynomial is greater than zero, this item is classified as positive, and negative otherwise. A high-degree polynomial can be wiggly, so it could match a certain set of training points well. However, if it is too wiggly, the classifier might make some mistakes at other points, and hence the polynomial is regarded with a high capacity. A simpler method is the threshold of a linear function that may not well fit the given set of training points because of its low capacity. The VC dimension is defined as follows: for a set of indicator functions, if there are

H samples that can be separated by a function of the function set in all possible 2^K forms, then the function set is able to break up the H samples. The VC dimension of the function set, which is the maximum number of samples H it can break up, is infinite if there are functions for any number of samples that can be broken up.

The VC dimension reflects the learning ability of the set of functions. The larger the VC dimension is, the more complex the learning machine (capacity greater) is; thus, it is a measure of the complexity of the learning machine. From another perspective, if the function f is used to represent a learning machine, a discriminant function EF is determined by a, and the VC dimension is the maximum number of training samples that the learning function learns all possible binary identifications that can be correctly given by its classification function. Unfortunately, there is no universal theory of computing VC dimensions for arbitrary function sets, and the VC dimension is known only for some special function sets. For example, the VC dimension of the linear classifier and the linear real function in N-dimensional space is $n+1$. Let us consider a simple example to further understand the VC dimension.

The set of instances X is a point (x, y) on a two-dimensional real plane, assuming that space H is all linear decision-making lines.

As shown in Figure 2.2, except for three points on the same line, the subsets of three points from x can be classified by any linear decision-making line, whereas the subsets of four points from x cannot be classified by any h in H, and hence $VC(H) = 3$.

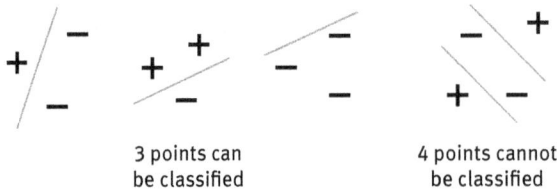

3 points can be classified   4 points cannot be classified

Figure 2.2: VC dimension diagram of the linear classifier.

The VC dimension measuring the complexity of a hypothetical space is not to use a different number of hypotheses $|H|$, but rather to use the number of different instances of X that can be thoroughly distinguished by H. The process is called splitting, which can be simply read as classification. The ability of H to split a set of instances is a metric of the ability to represent the notion of a goal defined on these instances. If any finitely large subset of X can be split by H, $VC(H) = \infty$, for any finite H, $VC(H) \leq \log_2 |H|$. Using the VC dimension as a metric of H complexity, it is possible to derive another solution to this problem, similar to the boundary of eq. (2.2), that is

$$m \geq \frac{1}{\varepsilon}\left(4\log_2\frac{2}{\delta} + 8VC(H)\log_2\frac{13}{\varepsilon}\right) \tag{2.3}$$

Equation (2.3) shows that to carry out PAC learning successfully, the required number of training samples should be proportional to the logarithm of $(1-\delta)$, proportional to $VC(H)$, and logarithmically proportional to $1/\varepsilon$.

3 Decision tree learning

Introduction

A decision tree is a flowchart structure where each internal node compares a "test" of an attribute. We consider an example of tossing a coin, where the probability of the flip coming up as a head or a tail is considered; each branch and each leaf node in a decision tree will denote a test result and a class label, respectively, that is taken for the decision based on the calculation of all attributes (M. W. H., 1969). In a decision tree, the path from the root to a leaf node represents a classification rule for decision analysis. Both a decision tree and its closely related influence map are used as visual and analytical decision-making tools that can determine the expected values or utility features from among competing alternatives. A decision tree consists of three categories of nodes in its diagram: a decision node – typically represented by a square; a chance node – typically denoted by a circle; and an end node – typically defined by a triangle. Decision trees are primarily used for studying and managing operations. For making decisions online without having to recall incomplete knowledge, the decision tree should be used along with the best probability model or the online selection model. Another example of application of decision trees is during descriptive ways of calculating conditional probabilities. To sum up, decision trees, influence maps, utility features, and other decision-making analysis tools and methods are used as examples when teaching students in schools of business, public health, and health economics different operations to understand the related scientific methods of research or management.

The decision tree is also an effective inductive reasoning method that can describe concept space. Each decision or event (i.e., the state of nature) can lead to two or more events and different output branches. Since these decision branches are drawn like the branches of a tree, the method is called a decision tree, which is always generated top-down (Kamiński, 2018). One advantage of the decision tree over a neural network is that a decision tree can generate rules, whereas neural networks fail to provide the appropriate reasons when making decisions. The learning methods based on a decision tree, which has been widely used in various fields, can be used for independent multi-concept learning because of being simple and quick.

3.1 Overview of a decision tree

Concept classification learning (CLS) began to develop in the 1960s. The CLS learning system, developed by Hunt, Marin, and Stone in 1966, is used to study individual concepts. J.R. Quinlan developed the Iterating Binomial 3 (ID3) algorithm in 1979, and further summarized and simplified it in 1983 and 1986, making it a model

https://doi.org/10.1515/9783110595567-004

for decision tree learning algorithms. Schlimme and Fisher transformed ID3 in 1986 to create a buffer at every possible decision tree node so that the decision tree can be incrementally generated, yielding the ID4 algorithm. In 1988, Utgoff proposed the ID5 learning algorithm based on ID4 to further improve its efficiency. In 1993, Quinlan further developed the ID3 algorithm, improving it into the C4.5 algorithm. The other kind of decision tree algorithm is the classification and regression tree (CART). Unlike C4.5, CART's decision trees are generated by binary logic problems. Each tree node has only two branches, including positive and negative examples of learning instances. The basic idea of the decision tree is to construct a tree with the fastest declines of entropy depending on information entropy as a measure. When the entropy value at the leaf nodes is zero, all the instances in each leaf node belong to the same class (Wikipedia, 2019).

Decision tree learning uses a top-down recursive approach. Each node of each layer is subdivided into sub-nodes depending on a certain attribute value, and the classified instance is compared with the attribute value related to the node at each node and is expanded to the corresponding sub-node depending on the comparison result. This process ends when it reaches the leaf node of the decision tree, at which point it is concluded. Each path from the root node to the leaf node corresponds to a reasonable rule, and the relationship between the various parts of the rule (conditions of each layer) is a conjunctive one. The entire decision tree is considered a disjunctive rule in the set concept.

The main advantage of the decision tree algorithm is that it is self-learning. In the learning process, users can learn with only well-labeled training samples, without needing to know enough background knowledge. In real applications, if the existing rules are broken, the program will ask the correct label to the users, depending on which it will then generate new branches and leaves, which can then be added into the decision tree.

3.1.1 Decision tree

The decision tree is a hierarchical data structure consisting of nodes (for saving information or knowledge) and branches (for connecting various nodes). The tree is a special case of a graph, which is a more general mathematical structure, such as a Bayesian network (Kamiński 2018).

The decision tree is a data structure that describes the classification process. Starting from the root node at the top, various classification principles are introduced into the moon. According to these classification principles, the data set of the root node is divided into sets. The constraint is satisfied and ends. Figure 3.1 shows a decision tree for determining the species of an animal.

Figure 3.1 is an example of a learning process of a decision tree. Each branch represents the possible value of an attribute of the instance, and the leaf node is

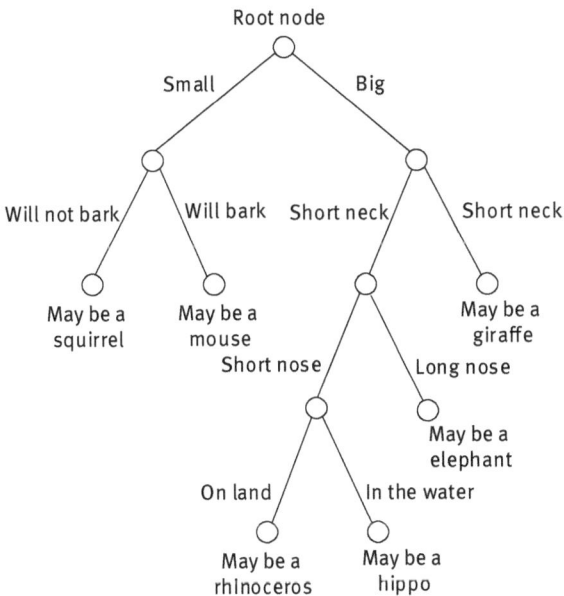

Figure 3.1: Decision tree example.

the final class. If the judgment is binary, then the path will be a binary tree. Every question in the tree falls to the next level of the tree. This type of tree is generally called CART.

The decision structure can be transformed into a production rule. It can be achieved by extensively searching the structure and generating "IF … THEN" rules at each node. The decision tree shown in Figure 3.1 can be converted into the following rules.

```
IF "big" THEN
        IF "short neck" THEN
                    IF "long nose" THEN May be an elephant
```

The decision process is formulated as big ∧ short neck ∧ long nose = may be an elephant.

When constructing a decision tree, we need deal with the following four issues.

(1) Collecting data to be classified, all attributes of which should be fully labeled.
(2) Designing a classification principle, that is, which attribute of the data can be used for classification and how to quantify the attribute.
(3) The choice of the classification principle, that is, among the many classification criteria, which criterion is chosen at each step makes the final tree more satisfying.

(4) Designing classification stop conditions. In practical applications, there are many attributes to represent data. The discriminative attributes are often limited. Therefore, the data set is split into subsets. Subsets should be made in such a way that each subset contains data with the same value for an attribute. The splitting process is repeated on each subset until leaf nodes are found that correspond to class labels in all the branches of the tree. The target (such as the entropy reduction criterion in ID3) does not need splitting because the depth of the tree is very large.

The general object function of splitting is the minimization of total entropy of the whole decision tree. When splitting a node, a criterion that minimizes the entropy is selected. This scheme first selects the criterion with the greatest possibility of classification.

3.1.2 Property

1. Evidence is represented by attribute value pairs
Evidence is represented by a fixed attribute and its value. If the attribute is temperature, its value is hot and cold. In the simplest case of learning, each attribute has a small amount of irrelevant value.

2. Objective function has discrete output values
The decision tree assigns a binary tree that can easily be expanded to more than two output values.

3. Need irrelevant description
The decision tree is in principle a representation of an irrelevant representation.

4. Tolerate errors in training data
There is strong robustness to errors in the training sample and the attribute values of the sample.

5. Training data can be missing values
The sample learning can have missing attribute values (not all samples).

3.1.3 Application

Decision trees have wide applications. For instance, patient classification with their condition, fault classification according to the cause, and loan applicant classification using payment credits are classification problems that classify input samples into the possible discrete sets.

3.1.4 Learning

First, let us introduce the Shannon information entropy (Luo, 2006).

1. Information

Let $p(a_i)$ denote the emitting probability of a_i from the source X. Before receiving the symbol a_i, the receiver's uncertainty of a_i is defined as the information $I(a_i)$ of a_i.

$$I(a_i) = -\log P(a_i)$$

2. Information entropy

Self-information can only reflect the symbolic uncertainty, and information entropy is used to measure the overall uncertainty of the entire source, defined as

$$H(X) = p(a_1)I(a_1) + p(a_2)I(a_2) + \cdots + p(a_r)I(a_r)$$

$$= -\sum_{i=1}^{r} p(a_i) \log p(a_i)$$

where r is all possible symbol types emitted by the source X. The entropy of information reflects the average amount of information of a source.

3.2 Design of decision tree

Let us consider an example of a fruit tree classification decision tree.

Fruit attributes can be described by the color, size, shape, and taste. For example, watermelon = Green $\wedge$ berg, apple = (green $\wedge$ medium size) $\vee$ (red $\wedge$ medium size). Its decision rules are shown in Figure 3.2.

3.2.1 Characteristics of decision trees

(1) The middle node corresponds to one attribute, and the branch under the node is the possible value of this attribute.
(2) Leaf nodes have a category tag, and each such node corresponds to a discriminant rule.
(3) Decision trees can produce both conjunctive and disjunctive rules.
(4) The rules generated by the decision tree are complete, and for any classification problem, they can be classified by constructing the corresponding decision tree.

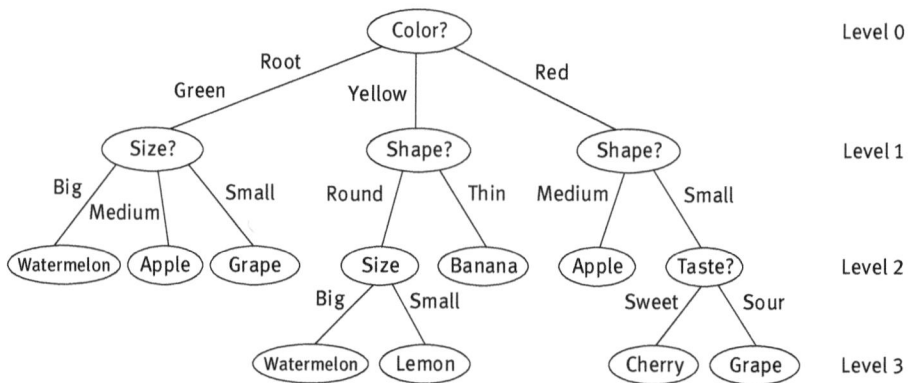

Figure 3.2: Decision tree example of fruit classification.

3.2.2 Generation of decision trees

Given a set of examples (sample set), a decision tree is generated to classify not only the existing samples in the set but also new samples. Let us consider the following example for understanding.

Wang is a manager of a famous tennis club. However, he was worried about the number of employees hired by the club. Some days it seems like everyone comes to the club to play tennis, making the staff so busy that they cannot handle the quantum of work. However, on other days, not many members turn out at the club. Since the reason for this is not known, the club ends up wasting a lot of money because the same number of employees still needs to be hired on those less-busy days.

Wang aims to determine when members will come in to play tennis based on the next weather forecast, and accordingly adjust the number of employees in a timely manner. So, first he must understand why people decide to play tennis.

The following records are available for two weeks: weather conditions are expressed as clear, cloudy, rainy; temperature is expressed in Chinese temperature; relative humidity is expressed as a percentage; wind conditions are expressed as windy or no wind. Correlating the weather data with customers visiting the club on those days, he eventually arrived at the data table shown in Table 3.1.

The decision tree model is built to solve the problem, as shown in Figure 3.3. The learning decision tree needs to solve the following questions:
(1) How to reduce the number of branches at the node?
(2) How to determine which node should test a property?
(3) When can a node become a leaf node?
(4) How to make an oversized tree smaller, that is, how to "prune" it?
(5) If the leaf node is still not pure, how to give it a category mark?
(6) How to deal with defective data?

First, the number of node branches is determined by two branches and multiple branches, as shown in the decision tree model.

Table 3.1: Tennis data.

Example	Weather	Temperature	Humidity	Wind power	Play tennis
1	Sunny	Hot	High	Weak	No
2	Sunny	Hot	High	Strong	No
3	Overcast	Hot	High	Weak	Yes
4	Rain	Mild	High	Weak	Yes
5	Rain	Cool	Normal	Weak	Yes
6	Rain	Cool	Normal	Strong	No
7	Overcast	Cool	Normal	Strong	Yes
8	Sunny	Mild	High	Weak	No
9	Sunny	Cool	Normal	Weak	Yes
10	Rain	Mild	Normal	Weak	Yes
11	Sunny	Mild	Normal	Strong	Yes
12	Overcast	Mild	High	Strong	Yes
13	Overcast	Hot	Normal	Weak	Yes
14	Rain	Mild	High	Strong	No

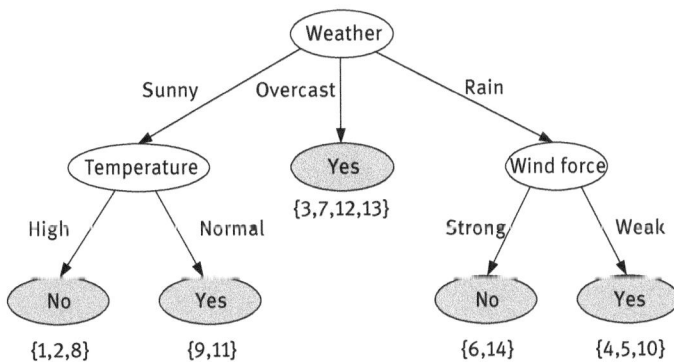

Figure 3.3: Decision tree model.

For (5), if the leaf node is still not "pure," that is, when there are multiple types of samples, this leaf node may be marked as the predominant sample type.

For (6), if some attributes of the sample to be identified are missing, when a node needs to detect this attribute, it can be discriminated on each branch.

(2) ~ (4) All three questions can be attributed to how to construct a "good" discriminant problem. We next discuss two algorithms – the ID3 algorithm and the C4 and C5 algorithms.

1. ID3 algorithm

The ID3 algorithm – an algorithm for decision trees proposed by Ross Quinlan in 1986 – is based on Occam's Razor: the simplest is the best one that can achieve the same goal. That is, simple models often correspond to strong generalization capabilities.

The ID3 algorithm is discussed in details next.

ID3 (Examples, Attributes), where Examples is a sample set and Attributes is a set of sample attributes.

1. Create Root;
2. if the element types in Examples are the same, it is a single-node tree, marked as the label of the category and returns Root;
3. if Attribute is empty, a single-node tree, labeled as the most common category label in Examples, returns Root;
4. the strongest classification capability attributes in A ← Attributes;
5. root decision attribute ← A;
6. the Examples of the elements according to the attributes of A are divided into several subsets, and the example$_i$ is a subset of attributes i;
7. if the example$_i$ is empty, add a leaf node under the new branch, the property marked the most common category in all examples; and
8. otherwise, add a child node ID3 (example$_i$, Attributes – {A}) under this branch.

As already discussed, "the most capable attribute classification," we use information gain to define the classification of attributes.

The entropy impurity of node N is defined as follows:

$$i(N) = - \sum_i p(\omega_j)\log_2 P(\omega_j)$$

where $P(\omega_j)$ is the frequency at node N belonging to the total number of ω_j samples.

Information gain of the attribute A at node N:

$$\Delta_A i(N) = i(N) - \sum_{v \in Value(A)} \frac{|N_v|}{|N|} i(N_v)$$

where *Value*(A) is a set of all possible values of the attribute *A*, N_v is a subset of attribute values *v* in *N*, and |*N*| is the number of elements in the set *N*.

Figure 3.4 shows an example of information gain calculation based on Table 3.1, node, attribute *A* = weather. Another example is given in Figure 3.5, attribute *A* = Outlook.

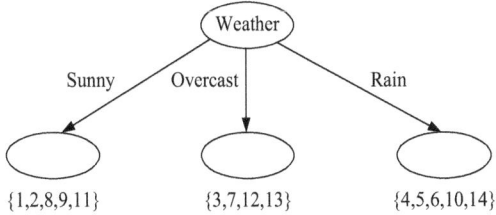

Figure 3.4: Example of information gain calculation based on Table 3.1.

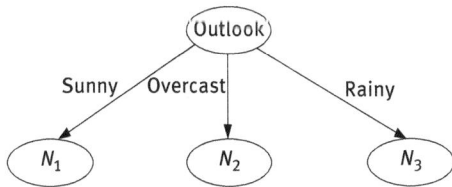

Figure 3.5: Decision tree based on Table 3.2.

$$i(N) = -\frac{9}{14}\log_2\frac{9}{14} - \frac{5}{14}\log_2\frac{5}{14} = 0.9403$$

$$\Delta_A i(N) = 0.9403 - \frac{5}{14}\left(-\frac{3}{5}\log_2\frac{3}{5} - \frac{2}{5}\log_2\frac{2}{5}\right) - \frac{4}{14}\left(-\frac{4}{4}\log_2\frac{4}{4} - \frac{0}{4}\log_2\frac{0}{4}\right)$$

$$-\frac{5}{14}\left(-\frac{2}{5}\log_2\frac{2}{5} - \frac{3}{5}\log_2\frac{3}{5}\right) = 0.246$$

At node *N*, select the test attribute based on the principle of maximum message gain:

$$\Delta_{weather}i(N) = 0.246$$

$$\Delta_{humidity}i(N) = 0.151$$

$$\Delta_{windpower}i(N) = 0.048$$

$$\Delta_{temperature}i(N) = 0.029$$

That is, choose the property of the weather to make the decision.

2. C4.5 algorithm

Since the ID3 algorithm does not have the "stop" and "pruning" technology, when the generated discriminant tree is relatively large, the data can be easily over-fit. In 1993, Quinlan added the "stop" and "prune" techniques to the ID3 algorithm and proposed the C4.5 algorithm to avoid over-fitting of the data.

1) Branch stopped

(i)   Verification Techniques: Part of the training samples is used as the verification set, and the node branches continue until the classification error for the verification set is the smallest.

(ii) Information gain threshold: Set the threshold β, stop the branch when the information gain is less than the threshold.

$$\max_s \Delta i(S) \leq \beta$$

(iii) Minimize global optimization: $a \cdots size + \sum_{Leaf\ node} i(N)$, size used to measure the complexity of the discriminant tree.

(iv) Hypothesis test.

2) Pruning

The growth of the discriminant tree is continued until the leaf nodes have the smallest impurity; then, for all leaf nodes that have a common parent node, consider whether they can be merged.

(i)   If the merger leaf nodes cause only a small increase in purity, the merger is performed.

(ii) Rule pruning: the first decision tree into the corresponding discriminant rules, and then trim the rule set.

So, what type of climate does it belong to?

Here is how to construct a decision tree that correctly classifies a training set using the ID3 algorithm from the training set given in Table 3.2.

When no weather information is given, based on the available data, we only know that the probability of playing on a new day is 9/14 and the probability of not playing is 5/14. The entropy at this moment is

$$-\frac{9}{14}\log_2\frac{9}{14} - \frac{5}{14}\log_2\frac{5}{14} = 0.940$$

There are four attributes, Outlook, Temperature, Humidity, and Windy. First, we need to decide which attribute is to be considered as the root node.

Statistics for each indicator: the number of times of playing and not playing at different values, as shown in Table 3.3.

Table 3.2: Climate training set.

No.	Attributes				Class
	Outlook	Temperature	Humidity	Windy	
1	Sunny	Hot	High	False	N
2	Sunny	Hot	High	True	N
3	Overcast	Hot	High	False	P
4	Rain	Mild	High	False	P
5	Rain	Cool	Normal	False	P
6	Rain	Cool	Normal	True	N
7	Overcast	Cool	Normal	True	P
8	Sunny	Mild	High	False	N
9	Sunny	Cool	Normal	False	P
10	Rain	Mild	Normal	False	P
11	Sunny	Mild	Normal	True	P
12	Overcast	Mild	High	True	P
13	Overcast	Hot	Normal	False	P
14	Rain	Mild	High	True	N

Table 3.3: Decision tree root node classification.

Outlook	yes	no	Temperature	yes	No	Humidity	yes	no	Windy	yes	no	Play yes	Play no
Sunny	2	3	Hot	2	2	High	3	4	False	6	2	9	5
Overcast	4	0	Mild	4	2	Normal	6	1	True	3	3		
Rainy	3	2	Cool	3	1								

The following shows how much the information entropy is when the value of the variable Outlook is known.

When Outlook = Sunny, the probability of playing is 2/5, and the probability of not playing is 3/5. Entropy = 0.971.

When Outlook = Overcast, Entropy = 0.

When Outlook = Rainy, Entropy = 0.971.

According to the statistics, Outlook has the probability values of 5/14, 4/14, and 5/14 for Sunny, Overcast, and Rainy, respectively; thus, when the value of Outlook is known, the entropy is $5/14 \times 0.971 + 4/14 \times 0 + 5/14 \times 0.971 = 0.694$.

In this case, the entropy of the system drops from 0.940 to 0.693, and the information Gain (Outlook) is $0.940 - 0.694 = 0.246$.

It is also possible to calculate Gain (Temperature) = 0.029, Gain (Humidity) = 0.151, Gain (Windy) = 0.048

Gain (Outlook) is with maximum value and thus the Outlook attribute will be selected in the step with information entropy dropped fastest.

After determining N_1, we take Temperature, Humidity, or Windy into consideration. Given Outlook = Sunny, a table similar to the one in Table 3.4 is developed based on historical data to calculate Gain (Temperature), Gain (Humidity), and Gain (Windy), with the largest being N_1 node Attributes.

Table 3.4: Decision tree N_1 node classification diagram.

Temperature	yes	no	Humidity	yes	no	Windy	yes	no	Play yes	no
Hot	0	2	High	0	3	False	1	2	2	3
Mild	1	1	Normal	2	0	True	1	1		
Cool	1	0								

Since Humidity divides the decision problem completely into yes and no, this node selects Humidity.

Similarly, node N_3 also draws a node classification map and chooses Windy as the decision attribute according to the information gain. Therefore, the final decision tree is as shown in Figure 3.6.

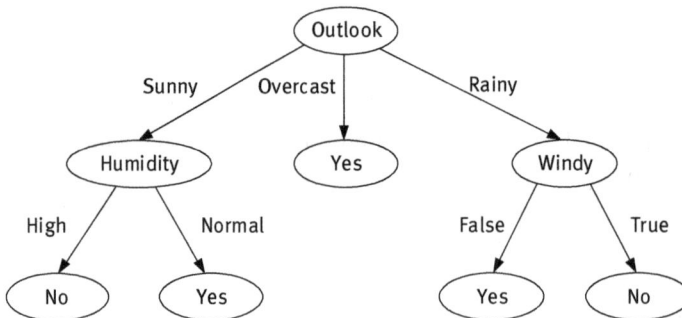

Figure 3.6: Decision tree based on Table 3.2.

Summary

This chapter mainly introduces the decision tree theory, including the basic concepts, the property, and applications. It emphasizes the algorithms needed to create a decision tree, such as ID3 and improved C4.5. ID3 tends to over-fit when the generated decision tree is too large; C4.5 is thus introduced for stopping and pruning functions to avoid over-fitting based on ID3. To better understand the algorithms of creating decision trees, we first briefly introduce the knowledge of information entropy using some specific examples to help readers understand and apply the decision tree.

4 Bayesian learning

Introduction

Bayesian decision theory is a basic method of statistical inference for pattern classification, in which the Bayes theorem is used to update the probability for a hypothesis as more evidence or information becomes available (Choudhuri, 2005) (Li, 2012; Albert, 2009). The basic principle lies in finding a compromise between the different classification decisions of probabilities and the corresponding decision costs. It is assumed that the decision problem can be described in terms of probability and that all relevant probability structures are known.

Before making the actual observations, it is assumed that judgments are made on the categories to be presented the next time, also assuming that any wrong decision will pay the same price and produce the same result. The only information that can be used is prior probability, which is ω_1 if $P(\omega_1) > P(\omega_2)$, and ω_2 otherwise.

If only one decision is required, then the decision rule discussed previously is reasonable. However, for multiple sentences, it might not be suitable to reapply such a rule since the same result will always be achieved.

For situations requiring multiple decisions, very little information is used to make the judgment. The example described next is used as an illustration to determine the male and female sex. The observed height and weight information can be used to improve the performance of the classifier since different sexes will produce different height and weight distributions and then represent them as probabilistic variables.

4.1 Bayesian learning

4.1.1 Bayesian formula

Suppose that x is a continuous random variable whose distribution depends on the state of the class and is expressed in the form of $P(x|\omega)$. It is called the class conditional probability density function of the class, that is, the probability density function of x at class ω, and the difference between $P(\omega_1)$ and $P(\omega_2)$ indicates the difference between the height of men and women.

Given the known prior probability $P(\omega_1)$, we also know the class conditional probability density $P(\omega_j)$, where $j = 1, 2$ and then assume that the height of a person is variable x, which affects the state of the class we are interested in. The union density in the category ω_j and with eigenvalue x can be written as $P(\omega_j, x) = P(\omega_j|x)P(x) = P(x|\omega_j)P(\omega_j)$, and this equation can be rearranged to

https://doi.org/10.1515/9783110595567-005

$$P(\omega_j|x) = \frac{P(x|\omega_j)P(\omega_j)}{P(x)}$$

This is the well-known Bayesian formula, where $P(\omega_j|x)$ is called the posterior probability, that is, x is known to belong to the category ω_j probability; $P(x|\omega_j)$ is the class conditional probability density (also known as the likelihood function); and $P(\omega_j)$ is the prior probability.

The Bayesian formula shows that by observing the value of x, the prior probability $P(\omega_j)$ can be transformed into the posterior probability $P(\omega_j|x)$, which is the probability that the category belongs to ω_j under the condition that the eigenvalue x is known. $P(x|\omega_j)$ denotes that under other conditions being equal, making it a larger value of ω_j is more likely to be the real category. Note that the posterior probability is mainly determined by the product of the prior probability and the likelihood function (Choudhuri, 2005).

4.1.2 Minimum error decision

If there is an observation x such that $P(\omega_1|x)$ is larger than $P(\omega_2|x)$, it is natural that we make a decision that the true category is ω_1. Similarly, if $P(\omega_2|x)$ is larger than $P(\omega_1|x)$, then ω_2 is more likely to be chosen. The following is the calculation of a decision, for a particular observation x, as

$$P(error|x) = \begin{cases} P(\omega_1|x) \text{ If it is judged as } \omega_2 \\ P(\omega_2|x) \text{ If it is judged as } \omega_1 \end{cases}$$

Obviously, for a given x, the decision can be made while minimizing the probability of error. Again, this rule can minimize the mean probability of error because the mean probability of error can be expressed as follows:

$$P(error) = \int_{-\infty}^{\infty} P(error, x)\mathrm{d}x = \int_{-\infty}^{\infty} P(error|x)P(x)\mathrm{d}x$$

In addition, if for any x, we guarantee that $P(error|x)$ is arbitrarily small, then the value of this integral will be arbitrarily small. This verifies the Bayesian decision rule under the condition of minimizing the error probability:

If $P(\omega_1|x) > P(\omega_2|x)$, judge as ω_1; otherwise, judge as ω_2.

According to the above-mentioned rules, the error probability of judgment can be written as

$$P(error|x) = \min[P(\omega_1|x), P(\omega_2|x)]$$

This form of judgment rule emphasizes the importance of posterior probability. By reusing the formula for calculating the posterior probability, one can transform this rule into a conditional probability and a priori probability of the form to describe. That is, to get the following completely equivalent decision rules:

$$\text{If } P(x|w)P(w_1) > P(x|w_2)P(w_2), \text{Discriminate as } w_1, \text{;otherwise as } w_2$$

This decision rule clearly shows that prior probability and likelihood probability are important for making a correct decision. The Bayesian decision rules combine them to achieve the minimum probability of error.

4.1.3 Normal probability density

In our experiments, we use the height and weight information of the human body, whose distribution is in accordance with the two-dimensional normal distribution. A brief introduction of the normal density thus follows.

The structure of a Bayesian classifier can be determined by the conditional probability density $P(x|w_1)$ and the prior probability $P(w_1)$. Of all the probabilities of all the studies, the most popular is the multivariate normal distribution.

The first is the continuous univariate normal density function:

$$P(x) = \frac{1}{\sqrt{2\pi}\delta} \exp\left[-\frac{1}{2}\left(\frac{x-\mu}{\delta}\right)^2\right]$$

From this probability density function, we can calculate the expected value and variance of x:

$$\mu = Ex = \int_{-\infty}^{\infty} x P(x) dx$$

$$\delta^2 = E(x-\mu)^2 = \int_{-\infty}^{\infty} (x-\mu)^2 P(x) dx$$

The univariate normalized density function is completely determined by two parameters: the mean μ and the variance δ^2. For brevity, it is usually abbreviated as $P(x) \sim N(\mu, \delta^2)$, which means that x follows a normal distribution with mean μ as the variance δ^2. Samples subject to normal distribution clustered around the mean, and the degree of dispersion was related to the standard deviation δ.

In practice, the more common case is the use of multidimensional density functions. The general form of d-dimensional multivariate normal density is as follows:

$$P(x) = \frac{1}{(2\pi)^{d/2}|\Sigma|^{1/2}} \exp\left[-\frac{1}{2}(x-\mu)^T \Sigma^{-1}(x-\mu)\right]$$

where x is a d-dimensional column vector, μ is the d-dimensional mean vector, Σ is the covariance matrix of $d \times d$, $|\Sigma|$ and Σ^{-1} are the values of the determinant and the inverse, respectively, and $(x - \mu)^T$ is the transpose of $(x - \mu)$. For the sake of simplicity, the same can be abbreviated as $P(x) \sim N(\mu, \Sigma)$.

Similarly, the mean and variance can be written as follows:

$$\mu = Ex = \int_{-\infty}^{\infty} x P(x) dx$$

$$\Sigma = E\left[(x - \mu)(x - \mu)^T\right] = \int_{-\infty}^{\infty} (x - \mu)(x - \mu)^T P(x) dx$$

The mean of a vector or matrix is obtained from the mean of its elements.

4.1.4 Maximum likelihood estimation

For practical problems, the conditional probability density of class conditions cannot be determined accurately and must be estimated in a certain way. For the example, the height and weight of a human body conform to a two-dimensional normal distribution, so we only need to estimate μ and Σ in the density function.

Parameter estimation is a classic problem in statistics, and some concrete solutions have been put forward, where maximum likelihood estimation is used. Assuming that there are n samples in the sample set D, $x_1, x_2, \cdots, x_n$ since the samples are extracted independently, the following equation can be obtained:

$$P(D|\theta) = \prod_{k=1}^{n} P(x_k|\theta)$$

We can think of $P(D|\theta)$ as a function of parameter vector θ, which is called the likelihood function under sample set D. By definition, the maximum likelihood estimate of the parameter vector θ is the vector of parameters $\hat{\theta}$ that maximizes $P(D|\theta)$. Moreover, the maximum likelihood estimation of the parameter vector θ is the one that best fits the existing set of observed samples.

To simplify the analysis and calculation, the logarithmic function of the likelihood function is usually used to find the desired parameter vector. The log-likelihood function is defined as follows:

$$I(\theta) = \ln P(D|\theta)$$

The target parameter vector $\hat{\theta}$ is a parameter vector capable of maximizing the log-likelihood function, that is,

$$\hat{\theta} = \arg\ \max I(\theta)$$

The following formula can be obtained:

$$I(\theta) = \sum_{k=1}^{n} P(x_k|\theta)$$

Derivation of the parameter vector θ in the previous equation yields

$$\nabla_\theta I = \sum_{k=1}^{n} \nabla_\theta P(x_k|\theta)$$

where $\nabla_\theta = \left[\frac{\partial}{\partial\theta_1}, \frac{\partial}{\partial\theta_2}, \ldots, \frac{\partial}{\partial\theta_P}\right]$.

In this way, if we let $\nabla_\theta = 0$, we can find the destination parameter vector $\hat{\theta}$.

For multivariate Gaussian functions, the mean is obtained using the maximum likelihood estimation method. The estimation of the variance is as follows:

$$\hat{\mu} = \frac{1}{n}\sum_{k=1}^{n} x_k$$

$$\hat{\sum} = \frac{1}{n}\sum_{k-1}^{n} (x-\mu)\,(x-\mu)^T$$

4.2 Naive Bayesian principle and application

4.2.1 Bayesian best hypothesis

The Bayesian best hypothesis is the most likely hypothesis given the data D and the prior probability of hypotheses H.

The Bayesian theorem is used to calculate the probability of hypotheses based on the prior probability of hypotheses, the observation probability under given hypotheses, and the observed data. $P(h)$ denotes the initial probability of h before training. $P(h)$ is often called the prior probability of hypothesis h, and it indicates the probability knowledge about which h is the correct hypothesis. Similarly, $P(D)$ denotes the prior probability of training data D. $P(h|D)$ represents the probability h given D, which is also known as the posterior probability of h. Using the Bayesian formula, one can calculate the posterior probability $P(h|D)$ as follows:

$$P(h|D) = \frac{P(D|h)\,P(h)}{P(D)}$$

In this formula, the data D represents a training sample of an objective function, and h is the candidate objective function space.

4.2.2 Naive Bayesian classification

The influence of Naive Bayesian classifiers on a given class is independent of other features, namely, the feature independence hypothesis.

For text categorization, it assumes that each word is independent of another. Figure 4.1 shows the principle.

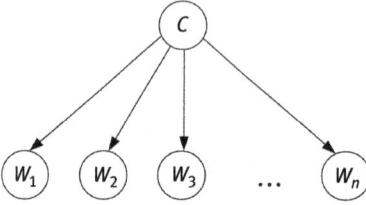

Figure 4.1: Naive Bayesian text classifier principle.

If the training sample set is divided into k classes, denoted $C = \{C_1, C_2, \ldots, C_k\}$, then each class C_i prior probability $P(C_i)$, $i = 1, 2, \ldots, k$. The number of samples whose value is C_i is divided by the number n of training samples. For the new sample d, the conditional probability of belonging to the class C_i is $P(d|C_i)$, as shown in eq. (4.1):

$$P(C_i|d) = \frac{P(d|C_i)P(C_i)}{P(d)} \tag{4.1}$$

$P(d)$ is a constant for all classes and can be ignored, and eq. (4.1) can be simplified as follows:

$$P(C_i|d) \propto P(d|C_i)P(C_i) \tag{4.2}$$

To avoid $P(C_i) = 0$, we use the Laplace probabilistic estimation and obtain eq. (4.3):

$$P(C_i|d) \propto P(d|C_i)P(C_i)$$

$$P(C_i) = \frac{1 + |D_{c_i}|}{|C| + |D_c|} \tag{4.3}$$

Here, $|C|$ represents the number of classes in the training set, $|D_{c_i}|$ represents the number of documents in the training set that belong to the class C_i, and $|D_c|$ represents the total number of documents included in the training set.

The Naive Bayes text classifier bases the unknown sample d on class C_i, as follows:

$$P(C_i|d) = \arg\ \max_j\{P(C_j|d)P(C_j)\},\ j = 1, 2, \ldots, k \tag{4.4}$$

The document d is represented by the words it contains, that is, $d = (\omega_1, \omega_2, \ldots, \omega_m)$, m is the number of feature words in d, ω_j is the j-th feature word, by the characteristic independence hypothesis, then eq. (4.5):

$$P(d|C_i) = P((\omega_1, \ldots, \omega_j, \ldots, \omega_m)|C_j) = \prod_{j=1}^{m} P(\omega_j|C_i) \qquad (4.5)$$

$P(\omega_j|C_i)$ represents the probability that the classifier predicts that word ω_j occurs in class C_i's document. Therefore, eq. (4.5) can be written as

$$P(d|C_i) \propto P(C_j) = \prod_{j=1}^{m} P(\omega_j|C_i) \qquad (4.6)$$

To avoid $P(\omega_j|C_i)$ being 0, the Laplace probabilistic estimation can be used.

4.2.3 Text classification based on Naive Bayes

With the evolution of the Internet over the recent years, the quantum of information available and accessible to people has grown exponentially. As a key technology for processing and organizing a large amount of text data, text categorization can solve the problem of information clutter to a large extent, making it convenient for users to accurately locate the required information. Text categorization has a wide range of application prospects as a technical basis for information filtering, information retrieval, search engine, database, digital library, and other fields.

Foreign automatic classification research began in the late 1950s when H.P. Luhn made groundbreaking research in this field. He first used the idea of word frequency statistics in text classification. In 1960, Maron published the first paper *On relevance probabilistic indexing and information retrieval* on automatic classification in the journal of ASM. In 1962, H. Borko et al. proposed the automatic classification of documents by factor analysis. Since then, many scholars have conducted fruitful research in this field. The research on foreign automatic classification can be divided into three stages: the first stage (1958–1964) was mainly the feasibility study of automatic classification; the second stage (1965–1974) was the experimental study of automatic classification; and the third stage (since 1975) was the automatic classification of the practical stage.

Some of the popular methods of text categorization in foreign countries include the Rocchi method and its variations, the k-nearest neighbor (kNN) method, decision tree, Naive Bayes, Bayesian network, and support vector machine (SVM). These methods have been extensively studied in the automatic classification of English and European languages, and numerous studies have shown KNN and SVM to be the best methods for English text classification. Numerous foreign researchers have studied the various problems in the field of English text categorization, and a number of comparative studies on several popular methods have been also carried out. Susan

Dumais and other scholars conducted a special comparative study of these five methods. Text classification refers to categorizing each text of a set of texts into a certain category automatically according to the content of the text based on a predefined classification system. The input of the system comprises a large number of texts that need to be classified, and the output of the system is the same as the text associated category. In short, text categorization is the labeling of documents with appropriate labels. Mathematically speaking, text classification is a mapping process that maps text that is not labeled into an existing category. The mapping can be either one-to-one mapping or one-to-many mapping because, usually, a piece of text can be associated with multiple categories.

The mapping rules of text classification are that the system summarizes the regularity of classification according to the data information of several samples in the known category and establishes the classification formula and the discrimination rules. On encountering a new text, the category to which the text belongs is determined based on the summarized category discrimination rules. In the theoretical research, far more studies on single-class classification have been carried out than on multi-class classification primarily because the single-class classification algorithm can be seamlessly converted into a multi-class one; however, this method assumes that all classes are independent and that there is no interdependence or other effects, which, in practical applications, in most cases, can be satisfied. Therefore, most of the experiments in text categorization are based on the study of single-class classification.

Naive Bayes Classifier (NBC) is introduced based on the Bayesian formula. The Bayesian classification algorithm – an important classification technology in data mining – can be compared with the classification algorithm, such as decision tree and neural network. Naive Bayesian classifier has a solid mathematical theory and the ability to integrate a priori information and data sample information, making it one of the trending topics in machine learning and data mining. Its simplicity and computational effectiveness have provided it with considerable robustness in practical applications, making it occupy a very important position in the field of text categorization.

1. Text classification basic concepts

In text classification, according to the predefined classification system, each text of the text collection is automatically classified into a certain category according to the content of the text. The input of the system is a large number of texts that need to be classified and processed, and the system output is the text associated with the category. In short, text categorization is the labeling of documents with appropriate labels. From a mathematical point of view, text classification is a mapping process that maps the text of an unidentified category into an existing category. The mapping can be either one-to-one mapping or one-to-many mapping because usually a text can be associated with multiple categories.

The mapping rule of text classification is that the system summarizes the regularity of the classification according to the data information of the ten samples in the known category and establishes the classification formula and the discrimination rule. On encountering a new text, the category to which it belongs is determined based on the summarized category discrimination rules.

In the theoretical research, far more studies on single-class classification have been carried out than on multi-class classification primarily because the single-class classification algorithm can be seamlessly converted into a multi-class one; however, this method assumes that all classes are independent and that there is no interdependence or other effects, which, in practical applications, in most cases, can be satisfied. Therefore, most of the experiments in text categorization are based on the study of single-class classification.

2. Text representation
Essentially, text is a string of characters that cannot be used by training algorithms to train or classify. To apply machine learning techniques to text classification problems, first the documents related to training and classification need to be converted into easily manageable vector forms of machine learning algorithms. That is, a variety of text representation methods, such as vector space model (VSM), the text of the document formalized representation. The popular VSM model proposed by G. Salton has good computability and operability. VSM was first successfully applied in the field of information retrieval and later has been widely used in the field of text classification.

VSM assumes that the category a document belongs to relates only to the frequency with which certain words or phrases appear in the document regardless of the position or order in which the words or phrases appear in the document. In other words, if the various semantic units (such as words and phrases) comprising the text are collectively referred to as "lexical items," and the frequency of appearance of the term in the text is called "frequency," then the word frequency information of each term contained in a document is sufficient for the correct classification.

In a VSM, the text is formalized as a vector in n-dimensional space:

$$D - < W_{term1}, W_{term2}, \ldots, W_{termn} >$$

where W_{term1} is the weight of the i-th feature. If the feature is selected as a word, then the importance of the word in representing the textual content is delineated.

This article uses Boolean weights. If the number of occurrences of a feature item is 0, its weight is 0; if the number of occurrences of a feature item is greater than 0, its weight is 1.

3. Feature extraction

In English text classification, the text is removed from stop words (is/are, on/at, etc. – words that have no effect on the meaning of the text), followed by words with the same root word and similar in meaning, such as work, working, worked. That is, when the word frequency is counted, the same word (e.g., work) is processed. However, the feature set is still with a high-dimensional space, which is too large for most classification algorithms. Therefore, it performs feature extraction based on document frequency (DF) to reduce the dimensionality of feature space and improve the efficiency and accuracy of classification.

The DF of a feature is the number of documents that have this feature in the document set. DF is the simplest feature extraction technique, which can be easily used for large-scale corpus statistics due to its linear computational complexity relative to the training corpus scale.

The use of DF as a feature choice is based on the basic assumption that terms with a DF below a certain threshold are low-frequency terms that contain no or less category information. Removing such terms from the original feature space not only reduces the dimensionality of the feature space, but also makes it possible to improve the classification accuracy.

4.3 Hidden Markov model and application

4.3.1 Markov property

In 1870, Russian organic chemist Vladimir V. Markovnikov first proposed the Markov model. If the "future" of a process relies on "the present" and does not rely on "the past," the process has Markovianity, or the process is a Markov process, $X(t+1) = f(X(t))$.

4.3.2 Markov chain

The Markov process with discrete time and state is called the Markov chain, denoted by $\{X_n = X(n), n = 0, 1, 2, \ldots\}$, and $X(n)$ is the result of observing the process of discrete states in time set $T_1 = \{0, 1, 2, \ldots\}$ successively.

The state space of hidden variables in Markov chain is recorded as $I = \{a_1, a_2, \ldots\}, a_i \in R$.

The conditional probability $P_{ij}(m, m+n) = P\{X_{m+n} = a_j | X_m = a_i\}$ is the Markov chain at time m, in state a_i, the transition probability of transitioning to state a_j at the time $m+n$.

4.3.3 Transition probability matrix

As shown in Figure 4.2, the weather transition probability matrix is shown in Table 4.1.

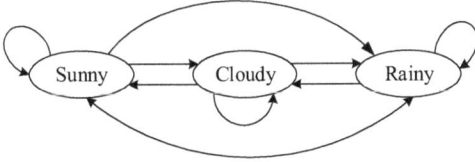

Figure 4.2: Weather transfer diagram.

Table 4.1: Weather transition probability matrix.

	Sunny	Cloudy	Rainy
Sunny	0.50	0.25	0.25
Cloudy	0.375	0.25	0.375
Rainy	0.25	0.125	0.625

Since the chain starts at any one of the states a_i at the moment m, it is inevitable that at the other moment $m+n$, the chain is shifted to any one of the states $a_1, a_2, \cdots$; thus, we have

$$\sum_{j=1}^{\infty} P_{ij}(m, m+n) = 1, \ i = 1, 2, \ldots$$

When $P_{ij}(m, m+n)$ is independent of m, the Markov chain is called a homogeneous Markov chain, and the Markov chain usually means a homogeneous Markov chain.

4.3.4 Hidden Markov model and application

1. An example of HMM

There are N cylinders, as shown in Figure 4.3. Each cylinder contains a large number of colored balls. The color of the ball is described by a set of probability distributions. The experiment is conducted as follows.
(1) According to the initial probability distribution, randomly select one of the N cylinders to start the experiment.
(2) According to the probability distribution of the ball color in the cylinder, a ball is selected at random, the color of the ball is recorded as O_1, and the ball is returned to the cylinder.

(3) According to the probability distribution describing the transfer of the cylinder, select the next cylinder randomly and repeat the above-mentioned steps.

Finally, a description of the ball's color sequence $O_1, O_2, \ldots$, called the observation
sequence O, is obtained.

In the above-mentioned experiment, the following points need attention:
(1) cannot directly observe the transfer between the cylinder;
(2) the color of the ball selected from the cylinder does not correspond to the cylinder;
and
(3) the choice of cylinder each time is determined by a set of transition probabilities.

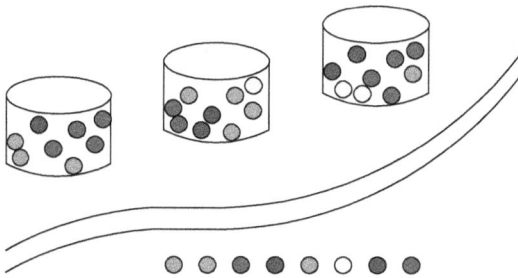

Figure 4.3: HMM example description.

2. HMM concept
The state of HMM is indeterminate or invisible and can only be manifested by a random process of observing the sequence (Chatzis, 2010). The observed events and states
do not correspond one by one, but rather through a set of probability distributions.

3. Composition of HMM
HMM is a two-stochastic process that has the following two components:
(1) the Markov chain, which describes the transfer of state, described by the transfer probability; and
(2) the general stochastic processes, which describes the relationship between the
state and the observed sequence, described by the probability of observation.

Figure 4.4 shows the HMM composition diagram.

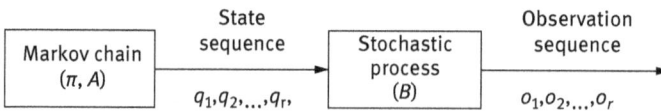

Figure 4.4: HMM composition diagram.

In the Markov process – a random process with no aftereffect – the probability of state at time t_m is only related to the state at time t_{m-1}, but not to the state before t_{m-1}, such as Brownian motion and Poisson process. Markov chains are time-discrete and discrete Markov processes.

Markov chains have two parameters: transition probability and initial probability. Between them, transfer probability $a_{kl} = P(\pi_i = 1 | \pi_{i-1} = k)$.

4. Basic algorithm of HMM

HMM mainly has three algorithms: the Viterbi algorithm, the forward–backward algorithm, and the Baum–Welch (B-W) algorithm.

1) Viterbi algorithm

(1) Using dynamic programming algorithm, the complexity is $O(K^2 L)$, where K and L are the number of states and sequence length, respectively.

(2) Initialize $(i = 0)$: $v_0(0) = 1, v_k(0) = 0, k > 0$.

Recursive $(i = 1 \ldots L)$: $v_l(i) = e_l(x_i) \max_k(v_k(i-1)a_{kl})$

$$ptr_i(l) = \arg\ \max_k(v_k(i-1)a_{kl})$$

Termination $p(x, \pi^*) = \max_k(v_k(L)a_{k0})$

$$\pi_L^* = \arg\max_k(v_k(L)a_{k0})$$

Backtracking $(i = 1 \ldots L)$: $\pi_{i-1}^* = ptr_i(\pi_i^*)$

2) Forward-backward algorithm

(1) Forward algorithm: dynamic programming, complexity with Viterbi.

Define the forward variable $f_k(i) = P(x_1 \cdots x_i, \pi_i = k)$.

Initialization $(i = 0)$: $f_0(0) = 1, f_k(0) = 0, k > 0$.

Recursive $(i = 1 \ldots L)$: $f_l(i) = e_l(x_i) \sum_k f_k(i-1)a_{kl}$

Termination: $P(x) = \sum_k f_k(L)a_{k0}$

(2) Backward algorithm: dynamic programming, complexity with Viterbi.

Define the backward variable $b_k(i) = P(x_{i+1} \ldots x_L | \pi_i - k)$.

Initialize $(i - L)$: $b_k(L) - a_{k0}$.

Recursive $(i = L - 1 \ldots 1)$: $b_k(i) = \sum_l a_{kl}e_l(x_{i+1})b_l(i+1)$.

Termination $P(x) = \sum_k a_{0l}e_l(x_1)b_1(1)$.

3) B-W algorithm

The revaluation equations are

$$A_{kl} = \sum_j \frac{1}{p(x^j)} \sum_i f_k^j(i) a_{kl} e_l(x_{i+1}^j) b_l^j(i+1)$$

$$E_k(b) = \sum_j \frac{1}{p(x^j)} \sum_{\{i|x_i^j = b\}} f_k^j(i) b_k^j(i)$$

5. HMM application

The main application of HMM is decoding (Chatzis, 2012). In biological sequence analysis, every value (observed value) in the sequence is used to infer which state it may belong to. There are two main solutions here: the Viterbi algorithm decoding and the forward–backward algorithm + Bayesian posterior probability.

1) Viterbi decoding

The result of the Viterbi algorithm is the best path, based on which a sequence of states corresponding to each observation can be derived directly.

2) Forward–backward algorithm + Bayesian posterior probability

Using Bayesian posterior probability, one can calculate the probability that a value in a sequence belongs to a state:

$$p(\pi_i = k|x) = \frac{P(x, \pi_i)}{P(x)}$$

$$p(x, \pi_i) = P(x_1 \ldots x_i, \pi_i = k) P(x_{i+1} \ldots x_L | \pi_i = k) = f_k(i) b_k(i)$$

The actual modeling process is as follows:
(1) according to the actual problem, determine the number of states and observation sequence;
(2) using several known sequences, estimate parameters using the B-W algorithm (the values of the transition probability a_{kl} and the output probability $e_k(b)$); and
(3) input unknown sequence with a Viterbi algorithm or Bayesian probability decoding.

Summary

The progress of modern society has led to the rapid development of a great variety of information, which, coupled with the rapid development of network resources, has led to the human society facing increasingly difficult information challenges. Increasing attention is being paid not only to the validity of information but also to the economy of access to information. Obtaining information conveniently and using it effectively have become the research hot spots of modern information technology. Text mining techniques such as text classification are one of the best ways to find and process information in all kinds of text carriers. In the current scenario where the information quantity is extremely large and information carrier is complex and fast-changing, it also provides a means to obtain effective information more economically.

5 Support vector machines

Introduction

As an important task of data mining, the purpose of classification is to learn a classification function or a classification model (classifier). Support vector machine (SVM) is a supervised learning method widely used in statistical classification and regression analysis (Cortes, 1995) (Hsu, 2002). SVMs were first proposed by Vapnik et al. in 1995. They have many unique advantages in solving small-sample, nonlinear, and high-dimensional pattern recognition and have been widely applied to face recognition, pedestrian detection, automatic text categorization, and other machine learning problems. The SVM is based on the VC dimension theory (refer to Chapter 2.2) and the structural risk minimization of the statistical learning theory. The SVM tries to find the best compromise between model complexity and learning ability based on limited sample information to obtain the best generalization.

5.1 Support vector machines

SVM is a machine learning method based on the statistical learning theory developed in the mid-1990s. It can improve the generalization ability of learning machines by structural risk minimization to minimize the risk of experience and the scope of confidence so as to obtain good statistical rules under the condition of small sample size. In general, the basic model of SVM is defined as a linear classifier with the maximum margin in the feature space. That is, the learning strategy involves maximizing the margin and finally solving a convex quadratic programming problem.

First, there are two types of linearly separable cases, as shown in Figure 5.1. The two types of training samples are solid points and hollow points. The optimal hyperplane of SVM is that the classification line can separate the two classes correctly (the training error rate is 0), and the classification margin is maximized. In Figure 5.1, H is the classification line that separates the two classes correctly. H_1 and H_2 are the parallel lines to H and pass through the point closest to H of the two classes. The margin is the vertical distance between H_1 and H_2.

Suppose the training data set is $(x_1, y_1), (x_2, y_2), \ldots, (x_n, y_n), x \in R^n, y \in \{+1, -1\}$. The linear discriminant function is

$$g(x) = \left(\omega^T x\right) + b \tag{5.1}$$

where $\omega^T x$ is the inner product of ω and x, and the equation of the hyperplane is $\left(\omega^T x\right) + b = 0$. Normalize the discriminant function, making all samples of the two classes satisfy $|g(x)| \geq 1$, where $g(x) \leq -1$ when $y = -1$, and $g(x) \geq 1$ when $y = 1$. The sample closest to the hyperplane is $|g(x)| = 1$.

https://doi.org/10.1515/9783110595567-006

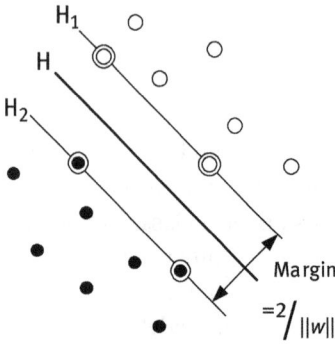

Figure 5.1: Optimal hyperplane.

The objective is to find the maximum margin hyperplane as the decision-making plane. First, we define the classification margin. When $g(x) > 0$, x is divided into ω_1 classes; when $g(x) < 0$, x is divided into ω_2 classes; and when $g(x) = 0$, it is the decision-making plane. Suppose x_1 and x_2 are two points on decision-making plane, then

$$\omega^T x_1 + b = \omega^T x_2 + b, \omega^T(x_1 - x_2) = 0 \tag{5.2}$$

As can be seen, ω and $x_1 - x_2$ are orthogonal and $x_1 - x_2$ is the direction of the decision-making plane, so ω is the normal vector of the decision-making plane. We express ω as

$$x = x_p + r\frac{\omega}{\|\omega\|} \tag{5.3}$$

where x_p is the projection vector of x on H, and r is the vertical distance from x to H. $\frac{\omega}{\|\omega\|}$ is the unit vector in the ω direction. Putting eq. (5.3) into eq. (5.1):

$$g(x) = \omega^T\left(x_p + r\frac{\omega}{\|\omega\|}\right) + \omega_0 = \omega^T x_p + \omega_0 + r\frac{\omega^T \omega}{\|\omega\|} = r\|\omega\| \tag{5.4}$$

Normalize r:

$$r = \frac{|g(x)|}{\|\omega\|} \tag{5.5}$$

From the above-mentioned analysis, the samples closest to the hyperplane satisfy $|g(x)| = 1$, so the classification margin is

$$margin = 2^* r = \frac{2}{\|\omega\|} \tag{5.6}$$

Therefore, to obtain the maximum of *margin* is to obtain the minimum of $\|\omega\|$ or $\|\omega\|_2$.

Because all training samples need to be correctly classified, the following condition needs to be fulfilled:

$$y_i[(\omega^T x) + b] - 1 \geq 0, \quad i = 1, 2, \ldots, n \tag{5.7}$$

to find the hyperplane making $\|\omega\|^2$ the minimum. The training samples on H_1, H_2 are found in eq. (5.7), which are called support vectors and marked with a circle in Figure 5.1. Therefore, the optimal hyperplane problem can be expressed as the following constraint optimization problem:

$$\text{minimize } \Phi(\omega) = \frac{1}{2}\|\omega\|^2 = \frac{1}{2}(\omega^T \omega) \tag{5.8}$$

The constraints are

$$y_i[(\omega^T x) + b] - 1 \geq 0 \quad i = 1, 2, \ldots, n$$

Constructing a Lagrange function as follows:

$$L(\omega, b, a) = \frac{1}{2}\|\omega\|^2 - \sum_{i=1}^{n} a_i(y_i.((x_i.\omega) + b) - 1) \tag{5.9}$$

where a_i is the Lagrange coefficient. Calculate the minimum of the Lagrange function for ω and b. Let the partial derivative of L with respect to ω be equal to 0:

$$\nabla_\omega L(\omega, b, a) = \omega - \sum a_i y_i x_i = 0$$

so that

$$\omega^* = \sum a_i y_i x_i \tag{5.10}$$

By substituting eq. (5.10) into L, we get the solution of L about ω:

$$L(\omega^*, b, a) = -\frac{1}{2}\sum_i \sum_j a_i a_j y_i y_j (x_i, x_j) - \sum_i a_i y_i b + \sum_i a_i \tag{5.11}$$

and partial derivative with respect to b:

$$\nabla_b L(\omega^*, b^*, a) = \sum a_i y_i = 0 \tag{5.12}$$

Substituting it into the solution of L to ω, we get the solution of L about ω and b:

$$L(\omega^*, b^*, a) = -\frac{1}{2}\sum_i \sum_j a_i a_j y_i y_j (x_i, x_j) + \sum_i a_i \tag{5.13}$$

Then, we solve the dual problem of the original problem. The dual of the original problem is as follows:

$$MaxQ(a) = -\frac{1}{2}\sum_{i}\sum_{j} a_i a_j y_i y_j (x_i, x_j) + \sum_{i} a_i$$

$$s.t. \sum_{i=1}^{n} a_i y_i = 0, \ a_i \geq 0, \ i = 1, \ ..., \ n \tag{5.14}$$

And we can have

$$\omega^* = \sum_{i=1}^{n} a_i y_i x_i \tag{5.15}$$

This is the quadratic optimization problem of the inequality constraint satisfying the Karush–Kuhn–Tucker (KKT) condition. In this way, ω^* and b^*, which can maximize eq. (5.15), need to be satisfied:

$$\sum_{i=1}^{n} a_i (y_i [(\omega.x) + b] - 1) = 0 \tag{5.16}$$

For most samples, they are not on the line closest to the classification hyperplane, namely, $y_i [(\omega.x) + b] - 1 > 0$. Thus, there must be a corresponding $a_i = 0$, that is, only the data points (support vectors) at the boundary are satisfied:

$$y_i [(\omega.x) + b] - 1 = 0$$
$$a_i \neq 0, \ i = 1, \ ..., \ n \tag{5.17}$$

They are only a small part of the overall sample and can significantly reduce the computational complexity to the original problem. Finally, the optimal classification function of the above-described problem is obtained:

$$f(x) = sgn\{(\omega^*. x) + b^*\} = sgn\left\{\sum a_i^* y_i (x_i.x) + b^*\right\} \tag{5.18}$$

Here, sgn() is a sign function. Since a_i of the non-support vectors is 0, the sum in the equation is only for the support vector. b^* can be obtained from any one of the support vectors using eq. (5.16). In this way, we obtain the SVM classifier in the two-class linearly separable cases.

However, not all two-class classification problems are linearly separable. For nonlinear problems, the SVM tries to transform it into a linear problem in another space by nonlinear transformation, in which the optimal linear hyperplane is solved. This nonlinear transformation can be achieved by an appropriate inner product function, namely, the kernel function. At present, the commonly used kernel functions mainly include the polynomial kernel, the radial basis kernel, and the sigmoid kernel. The parameters also have a great influence on the final classification result.

Previously, the new sample to be classified had to undergo a linear operation using ω and b first, and then it could be determined whether it is a positive or negative sample depending on whether the result is greater than 0 or less than 0.

Now with a_i, determining w is not necessary: only the inner product sum of all the samples in the new data and the training data needs to be calculated. From the KKT condition, $a_i > 0$ only for the support vectors and $a_i = 0$ for all other cases. Therefore, only calculating the inner product of the new sample and the support vector is enough.

The kernel function describes the core of SVM; it transforms the SVM into a nonlinear classification.

In Figure 5.2, we set all the points in the section between a and b on the horizontal axis as positive and the points on both sides as negative. Can we find a linear function to separate them? No, because the linear function in two-dimensional space refers to a straight line, and hence we cannot find a line that meets the conditions.

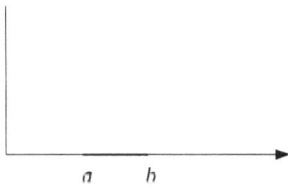

Figure 5.2: Two-dimensional linear inseparable example.

However, we can find a curve, such as the one shown in Figure 5.3.

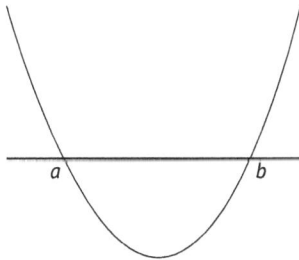

Figure 5.3: Example of a two-dimensional kernel function.

Obviously, determining whether the point in the curve is above or below can determine which class the point belongs to. This curve is known as the quadratic curve, and its function can be expressed as follows:

$$g(x) = c_0 + c_1 x + c_2 x^2$$

Thus, we first need to extend the feature x to three dimensions $(1, x, x^2)$ and then find the model between the feature and the result. This feature transform is called feature mapping. The mapping function is called Φ. In this example, $\Phi(x) = \begin{bmatrix} 1 \\ x \\ x^2 \end{bmatrix}$.

We want to apply the feature-mapped features instead of the original features to the SVM classification. In this way, the inner product in the previous equations needs to be mapped from $<x^{(i)}, x>$ to $<\Phi(x^{(i)}), \Phi(x)>$. In eq. (5.18), it can be seen that the linear classification uses the inner product of the original features $<x^{(i)}, x>$. In the nonlinear classification, we only need to use the inner product after the mapping. The choice of mapping depends on the characteristics of the sample and the classification effect.

However, for nonlinear classification, the feature mapping can increase the dimension rapidly, making it greatly challenging to calculate the speed. The kernel function solves this problem well. Next, we introduce the derivation.

Here, we formally define the kernel function. If the inner product of the original feature is $<x^{(i)}, x>$, and $<\Phi(x^{(i)}), \Phi(x)>$ is mapped, the kernel function can be defined as $K(x, z) = \Phi(x)^T \Phi(z)$. The meaning of this definition can be illustrated with an example as follows. Let $K(x, z) = (x^T z)^2$; we expand it as,

$$K(x, z) = (x^T z)^2 = \left(\sum_{i=1}^{m} x_i z_i \right) \left(\sum_{j=1}^{m} x_i z_i \right) = \sum_{i=1}^{m} \sum_{j=1}^{m} x_i y_j z_i z_j$$

$$= \sum_{i=1}^{m} \sum_{j=1}^{m} (x_i x_j)(z_i z_j) = \Phi(x)^T \Phi(z),$$

where Φ is.

$$\Phi(x) = \begin{bmatrix} x_1 x_1 \\ x_1 x_2 \\ x_1 x_3 \\ x_2 x_1 \\ x_2 x_2 \\ x_2 x_3 \\ x_3 x_1 \\ x_3 x_2 \\ x_3 x_3 \end{bmatrix}$$

In other words, the kernel function $K(x, z) = (x^T z)^2$ can only be equivalent to the inner product of the mapped features when such Φ is selected. In this case, the inner product of nine-dimensional vectors is represented by the kernel function of the three-dimensional vector, which greatly reduces the computational complexity.

There are many kernel functions. Mercer's theorem is used to determine the validity of kernel functions. It will not be introduced here. The common kernel functions are as follows.

Polynomial kernel: $K(x,z) = [x^T z + 1]$
Radial basis kernel: $K(x,z) = \exp(-\frac{|x-z|^2}{\sigma^2})$
Sigmoid kernel: $K(x,z) = \tanh(v(x^T z) + c)$

A summary of the kernel function is that different kernels use nonlinear combinations with different original features to fit a classification hyperplane.

There is another problem in SVM. We renew the linear classifier. When training a linear classifier with the minimum margin, if the sample is linearly separable, we can obtain the correct training result. However, if the sample is inseparable, the objective function has no solution and training failure will occur. In fact, this phenomenon is very common, so we introduce slack variables into SVM:

$$\Phi(\omega, \xi) = \frac{1}{2}\|\omega\|^2 + c\sum_{i=1}^{l}\xi$$

$$\omega^T x_i + b \geq +1 - \xi_i \quad y = +1$$

$$\omega^T + b \leq -1 + \xi_i \quad y_i = -1$$

$$\xi_i \geq 0 \quad \forall i$$

The value of c has a clear meaning: selecting a large value of means more emphasis on minimizing training errors.

5.2 Implementation algorithm

The usage of various inner product functions will lead to different SVM algorithms (Drucker, 1997). There are three main inner product functions related to the existing methods.

(1) Use the polynomial inner product function

$$K(x, x_i) = [(x \bullet x_i) + 1]^q \tag{5.19}$$

The SVM obtained in this case is a q-order polynomial classifier.

(2) Use the kernel-type inner product

$$K(x, x_i) = \exp\left\{-\frac{|x - x_i|^2}{\sigma^2}\right\} \tag{5.20}$$

The resulting SVM is a radial basis function (RBF) classifier. The basic difference between this method and the traditional RBF is that the center of each basis function corresponds to a support vector, and all of them and the output weights are automatically determined by the algorithm.

(3) Use the sigmoid function as the inner product

$$K(x, x_i) = \tanh(v(x \bullet x_i) + c) \tag{5.21}$$

Thus, the resulting SVM is a two-layer multilayer perceptron neural network, where not only the network weights but also the number of hidden layer nodes in the network is automatically determined by the algorithm.

As with many other conclusions in statistical learning theory, although SVM is proposed for classification, it can be generalized to the problem of continuous function fitting by defining an appropriate loss function.

At present, research on SVM mainly focuses on the comparison with some existing methods besides the theory research. For example, Bell Labs uses a comparative experiment conducted by the U.S. Postal Standard Handwritten Digits, which is a poorly identifiable database with every sample size being of 16×16 (256 dimensions). The training set has 7,300 samples, and the test set has 2,000 samples. Table 5.1 shows the test results of the classifier obtained by the manual method and several traditional methods, where the result of the two-layer neural network is to take the best of a two-layer neural network, and LetNet1 is a five-layer neural network designed specifically for this handwritten digit recognition problem. The experimental parameters and the results are shown in Table 5.2.

Table 5.1: Recognition results of the U.S. Postal Handwritten Database by the traditional method.

Classifier	Test error rate
Manual classification	2.5%
Decision tree method	16.2%
Two-layer neural network	5.9%
LetNet1	5.1%

Table 5.2: Experimental results of three kinds of SVM.

SVM type	Parameters in the function	Number of support vectors	Test error rate
Polynomial inner product	$q = 3$	274	4.0%
Radial basis inner product	$\sigma^2 = 0.3$	291	4.1%
Sigmoid inner product	$b = 2,\ c = 1$	254	4.2%

This experiment shows that the SVM has obvious advantages over the traditional methods and also shows that different SVM methods can achieve similar results (unlike the neural network, which relies heavily on model selection). In addition, three different SVMs are obtained in the experiment. The resulting support vectors are only a few of the total training samples, and more than 80% of the three sets of support vectors are coincident, which shows that the support vector is somewhat insensitive to different methods. Unfortunately, these attractive conclusions are currently observed only in limited experiments. If we can prove that they are correct, it will be a huge breakthrough in the theory and application of SVM.

In addition, there are some free software for SVM, such as LIBSVM, SVM[light], bsvm, mySVM, and MATLAB SVM Toolbox. Among them, LIBSVM is a simple, easy-to-use, fast, and effective SVM pattern recognition and regression software developed and designed by Associate Professor Lin from National Taiwan University. It provides not only a compiled executable file that can run on Windows but also the source code.

5.3 SVM example

A simple classification is performed using the SVM toolbox (made by Lin Zhiren, Taiwan), which can provide classification accuracy and support vector for each class. However, since the MATLAB toolbox cannot draw the classification hyperplane, we may use the training sample points as input to test the model. The test procedure and results are shown in Figures 5.4 and 5.5.

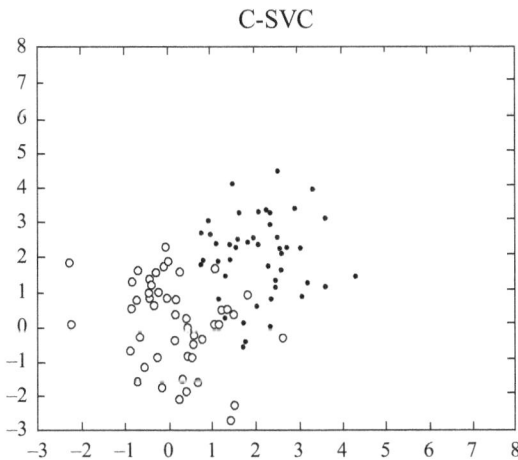

Figure 5.4: Training samples.

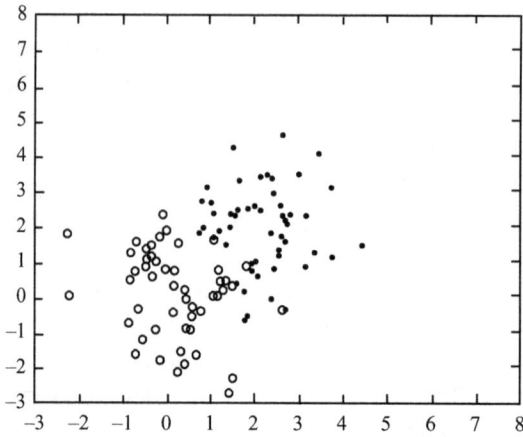

Figure 5.5: Test samples.

Classifier classification is as follows:

```
N=50;
n=2 * N;
x1=randn(2,N);
y1=ones(1,N);
x1=2+randn(2,N);
y2=-ones(1,N);
figure;
plot(x1(1, :), x1(2, :), 'o', x2(1, :), x2(2, :), 'k.');
axis([-3 8 -3 8])
title('C-SVC')
hold on;
X1=[x1, x2];
Y1=[y1, y2];
X=X1';
Y=Y1';
model=svmtrain(Y, X);
Y_later=svmpredict(Y, X, model);
%C1num=sum(Y_later>0);
%C2 num=2 * N-C1num;
%
%x3=zeros(2, C1num);
%x4=zeros(2, C2num);
```

```
figure;
for i=1:2 * N
    if Y_later(i) > 0
        plot(Xl(1, i), X1(2, 1), 'o');
        axis([-3 8 -3 8]);
        hold on
    else
        plot(Xl(1, i), X1(2, 1), 'k.');;
        hold on
    end
end
```

Furthermore, by promoting the optimal and generalized optimal classification hyperplane, we reach the following conclusions.

Theorem 5.1: If a group of training samples can be separated by an optimal classification hyperplane or a generalized optimal classification hyperplane, the expected upper bound for the test sample classification error rate is the ratio of the average support vector in the training samples to the total number of training samples.

$$E(P(error)) \leq \frac{E[SVM]}{Number\ of\ training\ samples - 1^\circ} \tag{5.22}$$

Therefore, the generalization of SVM is also independent of the dimension of the transformation space. By choosing a proper inner product function and constructing an optimal or generalized optimal classification hyperplane with few support vectors, we can obtain better generalization.

Here, the statistical learning theory uses a totally different approach from the traditional methods. Instead of reducing the dimension of the original input space (i.e., feature selection and feature transformation) as the traditional method does, statistical learning tries to increase the dimensions of the input space to make the input linearly separable (or nearly linearly separable) in high-dimensional space. Because only the inner-product operation is changed after the dimension increment, the complexity of the algorithm does not increase with the dimension increment, and the ability of generalization is not affected by the number of dimensions, hence making this method quite feasible.

5.4 Multi-class SVM

SVM was originally designed for two-class problems; thus, for dealing with multi-class problems, a suitable multi-class classifier is required. Currently, there are mainly two kinds of methods to construct SVM classifiers. The first one is the direct method, where all the classification decision functions are calculated simultaneously after properly changing the original optimization problem. Although this method seems simple, it has certain drawbacks, such as its computational complexity is relatively high, it is more difficult to achieve, and it is only suitable for small problems. The second method is the indirect method, which mainly uses the combination of a plurality of binary classifiers to construct multi-classifiers including the one-versus-rest method and the one-versus-one method.

(1) One-versus-rest (OVR SVMs). For each training, the samples of the specified class are grouped into one class, and the remaining samples are grouped into another class so that the samples of N classes construct N SVM classifiers. When classifying unknown samples, the class with the largest classification function value is taken as its belonging class.

(2) One-versus-one (OVO SVMs). This method is used to design an SVM classifier between any two classes, so a sample of N classes is needed to design SVM classifiers. When classifying unknown samples, the "voting" method is used, namely, the class with the maximum number of votes is the class of the unknown sample.

(3) One-versus-one (OVO SVMs). This method is used to design an SVM classifier between any two classes, so a sample of N classes is needed to design $N(N-1)/2$ SVM classifiers. When classifying unknown samples, the "voting" method is used, namely, the class with the maximum number of votes is the class of the unknown sample.

Summary

SVM considers statistical learning theory as its solid theoretical basis. It has numerous advantages, such as structural risk minimization, overcoming the problem of over-learning, and getting into the local minimum of the traditional method, besides having strong generalization ability. Using the kernel function, mapping to high-dimension space not only increases the computational complexity but also effectively overcomes the dimensionality disaster. The current SVM research also has certain limitations.

(1) The performance of SVM relies heavily on the choice of the kernel function, and there is no good way to help choose the kernel function for a particular problem.

(2) Training the speed and scale of SVM is another issue, especially for real-time control problems where speed is a major constraint on SVM. In response to this problem, Platt and Keerthi proposed the sequential minimization optimization (SMO) and the improved SMO method, respectively; however, they need further in-depth studies.

(3) The existing SVM theory only discusses the case of a fixed penalty coefficient, whereas, in fact, the two kinds of misclassification of positive and negative samples tend to cause different losses.

6 AdaBoost

Introduction

In machine learning, the decision tree is a prediction model, which represents a mapping between object attributes and object values. It is a predictive tree that relies on classification and training. Since its inception, it has been hailed as the classical algorithm in machine learning and data mining. There are several algorithms on the combination of the model ensemble (such as boosting, bagging, etc.) and decision tree; the objective of these algorithms is to generate N (may be more than a hundred) trees, which can greatly reduce the drawbacks of the single-decision tree (R., 2009) (Wang, 2008). Although each of these hundreds of decision trees is simple to a C4.5 single-decision tree, they are powerful combinations. In recent years, such as the ICCV 2009 Proceedings, there are many articles related to boosting and random forest. There are two basic types of algorithms that combine the model ensemble with decision trees: AdaBoost and random forest. AdaBoost is a typical representative of boosting, and random forest is a typical representative of bagging. Other relatively new algorithms are derived from the extension of both algorithms. Both single decision tree and model based derivative algorithms have both classification and regression applications. Here, we mainly introduce the basic principle, implementation, and application of AdaBoost and random forest, the two algorithms that are based on the classic decision tree.

6.1 AdaBoost and object detection

6.1.1 AdaBoost algorithm

The basic idea of AdaBoost is to use a large number of weak classifiers with general classification ability to form a strong classifier with strong classification ability as shown in the following formula:

$$F(x) = a_1 f_1(x) + a_2 f_2(x) + a_3 f_3(x) + \dots$$

where x is a vector; $f_1(x), f_2(x), f_3(x) \dots$ are weak classifiers; a_1, a_2, a_3 are weights; and $F(x)$ is a strong classifier.

As shown in Figure 6.1, some samples are classified in Figure 6.1(a). Every sample, that is, data point, has a class label and a weight, where black/dark represents the +1 class and gray/light represents the −1 class with a weight of 1. The straight line in Figure 6.1(b) represents a simple binary classifier. In Figure 6.1(c), a binary classifier with the lowest error rate is obtained by adjusting the threshold. This weak classifier has better classification ability than random classification. In Figure 6.1(d), the weight of the sample is updated, that is, the weight of the sample to be misclassified

https://doi.org/10.1515/9783110595567-007

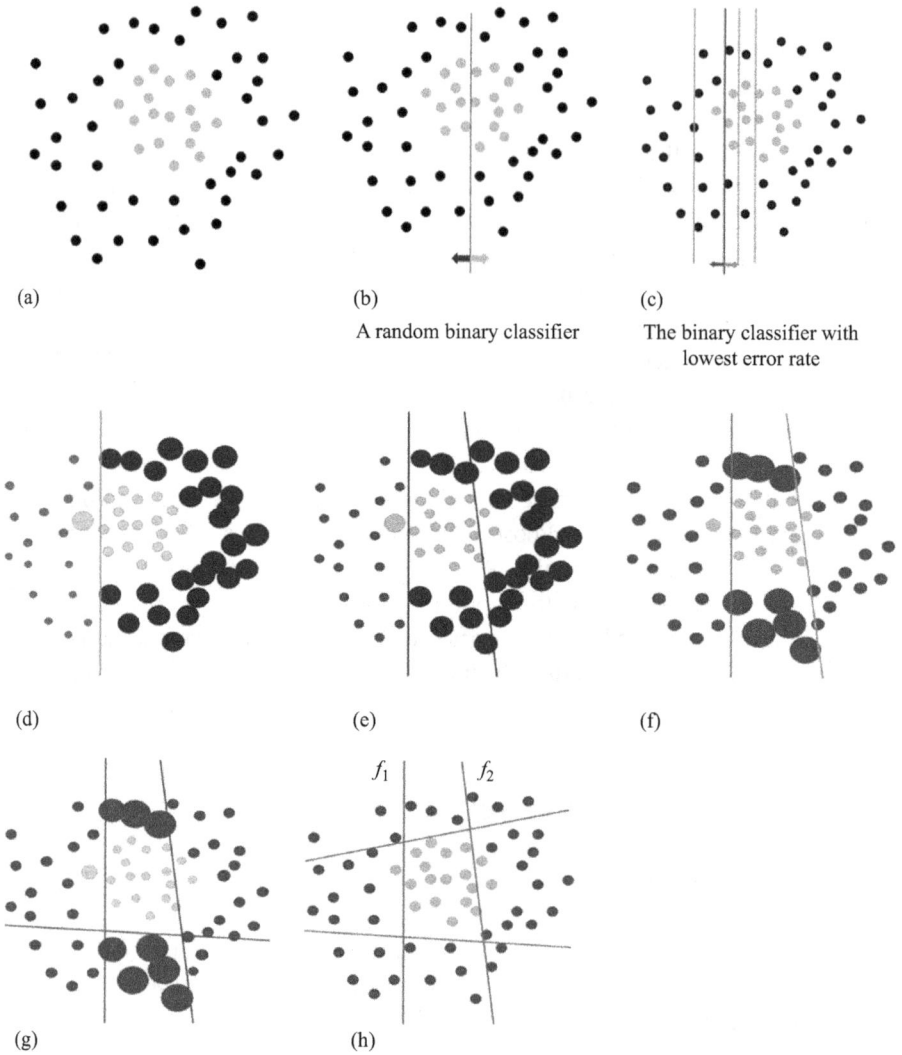

Figure 6.1: A strong classifier example.
Note: http://people.csail.mit.edu/torralba/shortCourseRLOC/boosting/boosting.html

is increased, so that a new data distribution is obtained. In Figure 6.1(e)–(h), according to the new data distribution, we look for the binary classifier with the lowest error rate. By repeating the above-mentioned process and adding the weak classifier to the strong classifier, we can get the strong classifier, shown in Figure 6.1(h). This strong linear classifier is one that includes weak linear classifiers in parallel.

The algorithm has been described next.

There are training sets of n training samples (x_1, y_1), (x_2, y_2), ..., (x_n, y_n), where $y_i = \{-1, +1\}(i = 1, 2, ..., n)$ corresponds to the sample's true or false. There are totally M negative samples and L positive samples. There are K simple features in the objects to be classified, which are expressed as $f_j(\bullet)$, $1 \leq j \leq K$. For the $i-th$ sample x_i, its K eigenvalues are $\{f_1(x_i), f_2(x_i), ..., f_k(x_i)\}$. For each eigenvalue f_j of the input feature, there is a simple binary classifier. The weak classifier for the $j-th$ feature consists of a threshold value θ_j, an eigenvalue f_j, and a bias p_j (only +1 or −1) indicating the direction of the inequality.

$$h_j = \begin{cases} 1, & p_j f_j \leq p_j \theta_j \\ -1, & others \end{cases}$$

Here, $h_j = 1$ indicates that the j-th feature determines that the sample is a true sample; otherwise, it is judged as a false sample.

The training target is to analyze the true and false samples obtained from the judgment, select the T weak classifiers with the lowest classification error rate, and finally combine them into a strong classifier.

The training method is as follows. For n given samples (x_1, y_1), (x_2, y_2), ..., (x_n, y_n), where $x_i \in X$, $y_i \in Y - \{-1, +1\}$, there are M negative samples and L positive samples in n samples.

6.1.2 AdaBoost initialization

Let $D_{t,i}$ be the error weight of the $i-th$ sample in the $t-th$ cycle, and initialize the error weights of the training samples according to the following formula: for the sample with $y_i = -1$, $D_{1,i} = 1/2M$; for the sample with $y_i = 1$, $D_{1,i} = 1/(2L)$. For $t = 1$, ..., T,

(1) the value is normalized such that $D_{t,i}$ is $D_{t,i} \leftarrow D_{t,i} / \sum_{j=1}^{n} D_{t,j}$ and $D_{t,i}$ is a probability distribution;

(2) for each feature j, train its weak classifier h_j, that is, determine the threshold value θ_j and the bias p_j such that the loss function $\varepsilon_j = \sum_{i=1}^{n} D_{t,i}|h_j(x_i) - y_i|$ of feature j reaches the minimum in this cycle;

(3) from all the weak classifiers identified in (2), find a weak classifier h_t with the smallest loss function, whose loss function is $\varepsilon_t = P_{r_i - D_i}[h_j(x_i \neq y_i)] = \sum_{i=1}^{n} D_{t,i}|h_j(x_i) - y_i|$, and add the weak classifier h_t to the strong classifier.

(4) update the weight of every sample, $D_{t+1,i} = D_{t,i}\beta_t^{1-e_i}$; the method for determining e_i is that the $i-th$ sample x_i is correctly classified, then $e_i = 0$; otherwise, $e_i = 1$. $\beta_t = \varepsilon_t/(1 - \varepsilon_t)$.

After T-round training, we can achieve a strong classifier formed by T weak classifiers in parallel:

$$H_{final}(x) = \text{sgn}\left(\sum_{t=1}^{T} a_t h_t(x)\right) = \begin{cases} 1, & \sum_{t=1}^{T} a_t h_t(x) \geq 0.5 \sum_{t=1}^{T} a_t \\ -1, & others \end{cases}$$

where $a_t = \log(1/\beta_t)$.

Here, α_t is a weak hypothesis, $H_{final}(x)$ is a final hypothesis, ε_t is the training loss of h_t, and D_t is the probability distribution of h_t.

The weak learning machine's job is to find a weak learning hypothesis h_t that is appropriate for the probability distribution D_t. The fitness of the weak hypothesis is measured by its error $\varepsilon_t = P_{r_i \sim D_i}[h_j(x_i \neq y_i)]$. The error ε_t is related to the distribution D_t of the weak learning machine. In practice, weak learning machines are calculated from the weights D_t on the training samples.

Once the weak learning hypothesis h_t holds, AdaBoost selects one parameter β_t. β_t is directly related to α_t, and α_t is the weight of h_t. $\varepsilon_t \leq 1/2$, then $\frac{\varepsilon_t}{1-\varepsilon_t} < 1$, $\beta_t < 1$, that is, the weight of the correctly classified sample becomes smaller, and the smaller the ε_t, the smaller the β_t; $\alpha_t \geq 0$, and the smaller the ε_t, the larger the α_t.

When initialized, all weights are set equal. After each cycle, the weight of the sample is redistributed, the weight of the sample misclassified is increased, and the weight of the sample correctly classified is reduced. The aim is to focus on the misclassified samples from the previous level. Strong classifiers are voted by a linear combination of the weights of every weak classifier and are ultimately determined by a threshold.

$$H_{final} = \begin{cases} 1, & \text{sgn}\left(\sum_{t=1}^{T} a_t h_t(x)\right) > \theta \\ -1, & others \end{cases}$$

Write ε_t as $1/2 - \gamma_t$, then the training error $(H_{final}(x)) \leq \exp\left(-2\sum_t \gamma_t^2\right)$, so if $\forall t : \gamma_t \geq \gamma > 0$, then the training error $(H_{final}(x)) \leq e^{-2\gamma^2 T}$. AdaBoost is more applicable because it does not require prior knowledge of γ or T and can make $\gamma_t \gg \gamma$.

We are, however, not concerned with the error of the training set, but the error of the test set. Then, with the increase in the number of training, will there be overfitting? Is not as Occam's razor said, the simple is the best?

The result in the classic method is shown in Figure 6.2, which is the result of boosting C4.5 on the "letter" data set.

Figure 6.2 shows that the error rate of the test set does not increase as training runs, even after 1,000 rounds. The test set error rate keeps decreasing even when the training error is 0.

Is AdaBoost always maximizing the classification margin? No. The classification margin trained by AdaBoost may be significantly less than the maximum (R,

Daubechies, Schapire 04). If we finally train a simpler classifier, is it possible to compress it, or can we not obtain a simple classifier by boosting?

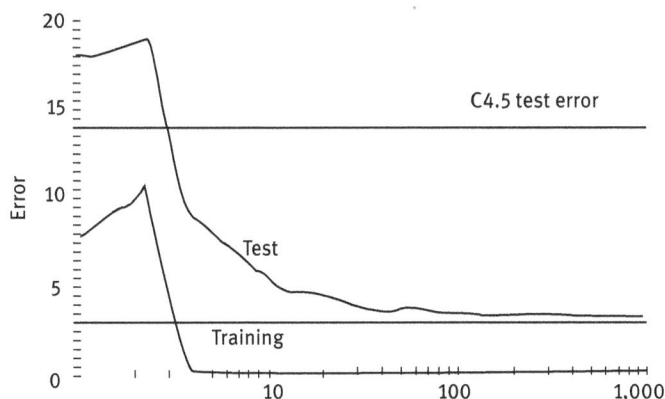

Figure 6.2: A typical example.

Consider the XOR problem shown in Figure 6.3, where x_1 and x_2 are the values of the first dimension and the second dimension of sample x, respectively. As shown in Figure 4.4, $h_1(x)\tilde{h}_8(x)$ are eight simple binary classifiers.

$$\begin{cases} (x_1 = (0, +1), \; y_1 = +1) \\ (x_2 = (0, -1), \; y_2 = +1) \\ (x_3 = (+1, 0), \; y_3 = -1) \\ (x_4 = (-1, 0), \; y_4 = -1) \end{cases}$$

Figure 6.3: XOR problem.

$$h_1(x) = \begin{cases} +1, & x_1 > -0.5 \\ -1, & \text{others} \end{cases} \qquad h_2(x) = \begin{cases} -1, & x_1 > -0.5 \\ +1, & \text{others} \end{cases}$$

$$h_3(x) = \begin{cases} +1, & x_1 > +0.5 \\ -1, & \text{others} \end{cases} \qquad h_4(x) = \begin{cases} -1, & x_1 > +0.5 \\ +1, & \text{others} \end{cases}$$

$$h_5(x) = \begin{cases} +1, & x_2 > -0.5 \\ -1, & \text{others} \end{cases} \qquad h_6(x) = \begin{cases} -1, & x_2 > -0.5 \\ +1, & \text{others} \end{cases}$$

$$h_7(x) = \begin{cases} +1, & x_2 > +0.5 \\ -1, & \text{others} \end{cases} \qquad h_1(x) = \begin{cases} -1, & x_2 > +0.5 \\ +1, & \text{others} \end{cases}$$

Figure 6.4: Simple binary classifier.

Let us now discuss how AdaBoost trains the strong classifiers.

(1) The first step is to invoke the basic learning rule based on the initial data set, namely, the simple binary classifier. h_2, h_3, h_5, and h_8 have a classification error of 0.25. We assume that h_2 is chosen as the first classifier. Thus, x_1 is misclassified, that is, the error rate is $1/4 = 0.25$. The weight of h_2 is 0.5.

(2) The weight of sample x_1 increases, and the simple binary classifier is called again. At this time, h_3, h_5, and h_8 have the same classification error. Assuming h_3 is chosen, we obtain the weight of 0.80.

(3) The weight of sample x_3 increases, and only h_5 and h_8 have the lowest error rate at the same time. Assuming h_5 is chosen, we obtain its weight of 1.10. The weak classifier obtained from the above-mentioned steps and its weight can be voted out of a strong classifier, and then the formed strong classifier can correctly classify all the samples. In this way, AdaBoost can train a strong classifier with a nonlinear zero-error rate by combining weak imperfect linear classifiers.

6.2 Robust real-time object detection

6.2.1 Rectangular feature selection

As shown in Figure 6.5, there are mainly three types of rectangle features.

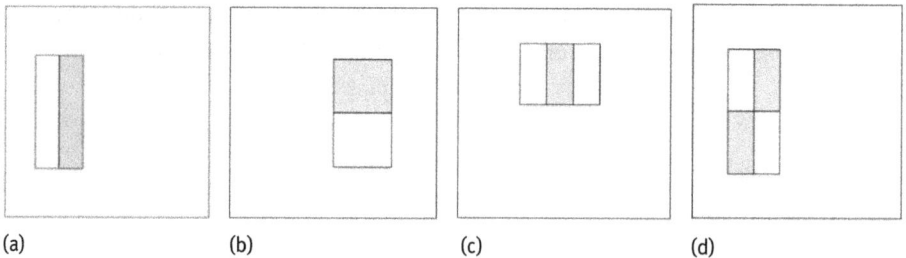

(a)   (b)   (c)   (d)

Figure 6.5: Rectangular features.

The two-rectangular features, shown in Figure 6.5(a) and Figure 6.5(b), are divided into left and right structures and up and down structures to represent the edge information.

The three-rectangular features, as shown in Figure 6.5(c), are divided into the left–central–right structure and the upper–central–lower structure to represent line information.

The four-rectangular feature, as shown in Figure 6.5(d), is a diagonal structure of four rectangles, indicating oblique boundary information.

In a 24 × 24 basic detection window, the number of different types of features and different scales of features can reach up to 49,396. When selecting classification features, considering the real-time requirements of computer recognition, the feature selection should be as simple as possible, the feature structure should not be too complicated, and the computational cost should be small. In contrast to

more expressive, easy-to-manipulate filters, the motivation behind using rectangular features is its powerful computational efficiency.

6.2.2 Integral image

Defining the gray level of every pixel of every image as $i(x,y)$, every pixel value
$ii(x,y)$ in the integral image of the image is expressed as

$$ii(x, y) = \sum_{x' \le x, y' \le y} i(x', y')$$

that is, the integral graph value of the point (x,y) in Figure 6.6 is the pixel gray
value summation of the gray rectangular area.

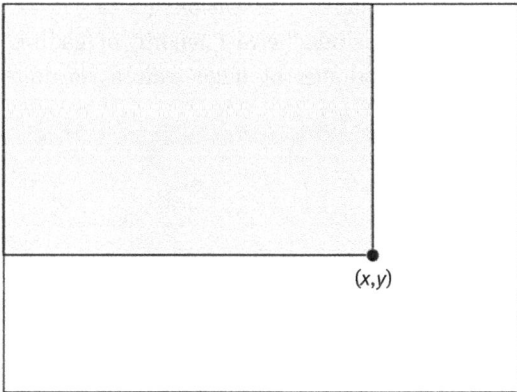

(x,y)

Figure 6.6: Example of the integral image value (x,y).

For the integral map value of any point in an image, we can accumulate the rows and
columns by one iteration.

$$s(x, y) = s(x, y - 1) + i(x, y)$$
$$ii(x, y) = ii(x - 1, y) + s(x, y)$$

where $s(x,y)$ is the column integral of the point (x,y) but does not contain the value
of (x,y). At the beginning of the iteration, $s(x, -1) = 0$, $ii(-1, y) = 0$. The integral
graph can easily sum the gray values in any rectangle in the image.

For example, we can use $ii(4) + ii(1) - ii(2) - ii(3)$ to sum the gray values of the
rectangle D in Figure 6.7. Using six, eight, and nine corresponding reference areas, one
can calculate the two-rectangular, three-rectangular, and four-rectangular features
easily.

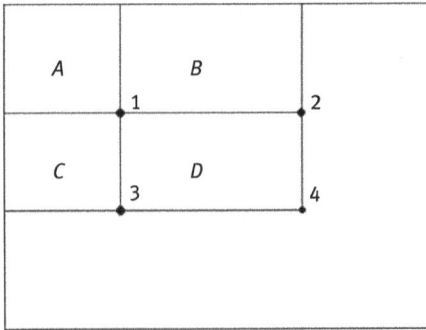

Figure 6.7: Summation of pixel gray values.

Given a set of features and a picture training set with class labels, a variety of machine learning methods can be applied. However, since there are 45,396 features in each picture window, the computation of all the features is unfeasible, making it necessary to use a fairly greedy learning algorithm to exclude the vast majority of features. Choosing fewer effective features from a large number of huge features is quite challenging.

6.2.3 Training result

Figure 6.8 shows that when the low false correct rate approaches 10^{-4}, the detection rate can reach 0.95. AdaBoost's training process selects the less-effective feature from a large number of huge features. For face detection, the rectangular feature initially selected by AdaBoost is crucial and has a physical meaning.

Figure 6.9 shows the first and second features obtained through Viola et al.'s learning process. The first feature indicates the horizontal area of the human eye, which is darker than the gray area of the upper cheek region. The second feature is used to distinguish the light and dark borders between the human eye and the nasal area.

By constantly changing the threshold of the final classifier, a binary classifier can be constructed with a detection rate of 1 and a false correct rate of 0.4.

6.2.4 Cascade

The strong classifiers are connected in series to form a hierarchical classifier. The threshold of strong classifiers of each layer is adjusted so that each layer can pass through almost all of the true samples and reject a large part of the false samples. Moreover, since the previous layers use a very small number of rectangular features

Cascading classifier and single classifier Roc curve recognition

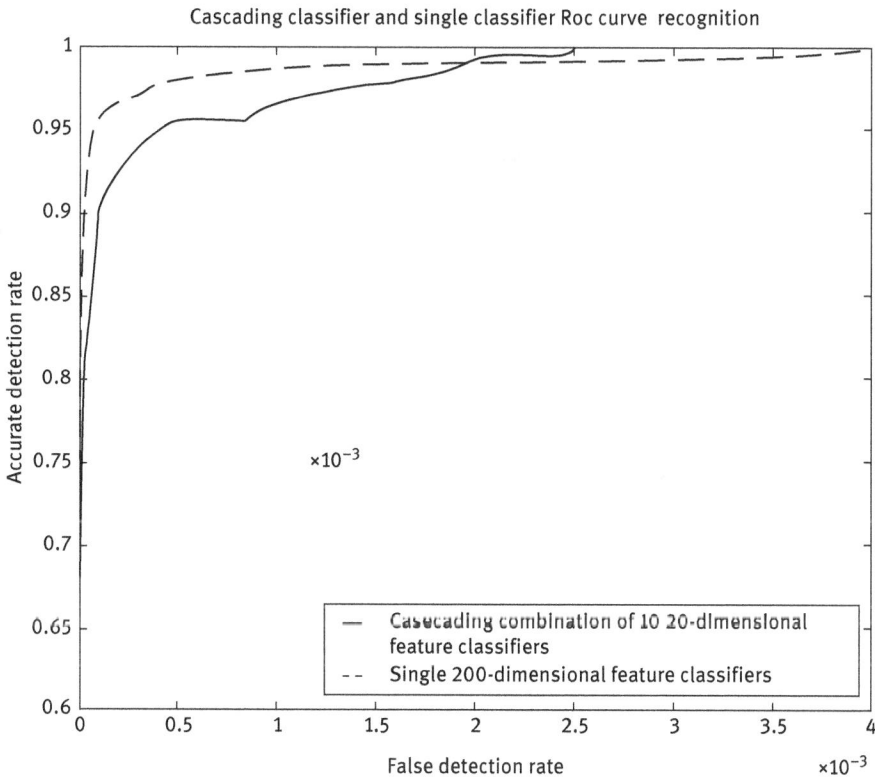

Figure 6.8: Receiver operating characteristic (ROC) curve of AdaBoost classifiers and a classifier.

and calculate very fast, the fewer candidate matching images pass later. That is, we should follow the idea of "First Heavy Later Light" in tandem and place the strong classifiers with the simpler structure and more important features first, so that a large number of false samples can be excluded first. With advances made in technology, the number of rectangular features keeps increasing; however, the amount of computation is now reduced and the speed of detection is increasing, giving the system a very good real time as shown in Figure 6.10.

A total of 4,916 experimental face samples were selected from the existing face database, which were manually cut, balanced, and normalized into basic 24×24 pictures; then, 1,000 negative samples from 9,500 faces without human faces were randomly selected in the picture.

The resulting detector has 32 layers and 4,297 features, as listed in Table 6.1.

The speed of the detector depends on the number of features. On the MIT-CMU test set, the average number of features calculated for each window is eight out of 4297 features. In a common personal computer, the time needed to process a 384×384 picture is 0.067 s.

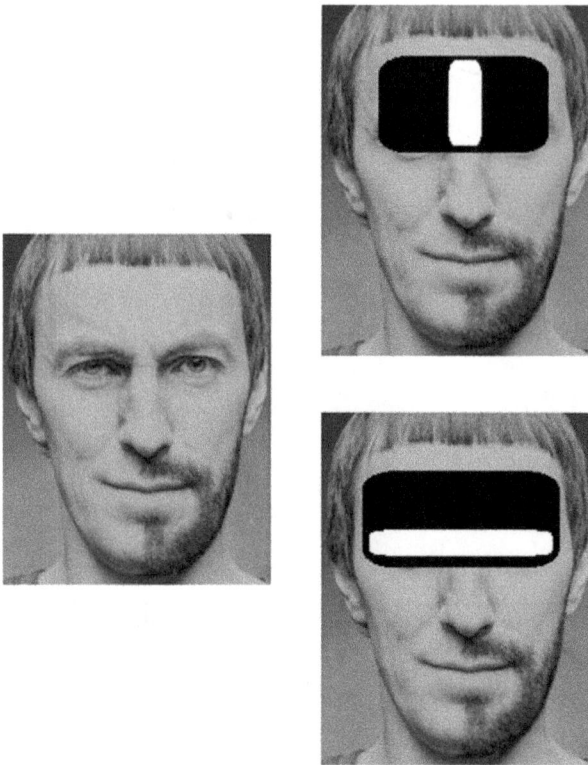

Figure 6.9: Features selected by AdaBoost.

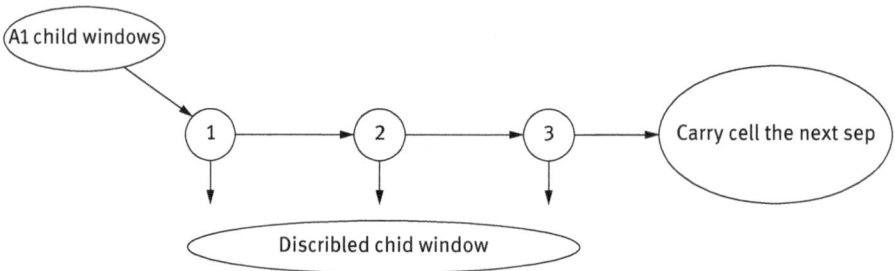

Figure 6.10: Cascade test schematic.

The test set uses the MIT + CMU positive face training set containing 130 images and 507 marked positive faces. The results in Table 6.2 are compared with the detections of several of the best face detectors.

Table 6.1: Results (Elgammal, 2005).

Layer number	1	2	3~5	6,7	8~12	13~32
Number of features	2	5	20	50	100	200
Detection rate	100%	100%	–	–	–	–
Rejection rate	60%	80%	–	–	–	–

Table 6.2: Comparison of classification ability (Elgammal, 2005).

False alarm rate	10	31	50	65	78	95	110	167	422
Viola–Jones		78.3%	85.2%	88.8%	89.8%	90.1%	90.8%	91.1%	91.8%	93.7%
Rowley–Baluja Kanade	83.2%	86.0%	–	–	–	89.2%	–	90.1%	89.9%
Schneiderman–Kanade	–	–	–	94.4%	–	–	–	–	–
Roth–Yang–Ajuha	–	–	–	–	94.8%	–			

6.3 Object detection using statistical learning theory

The basic idea of AdaBoost is to superpose a large number of weak classifiers with general classification ability through a certain method to form a strong classifier with strong classification ability. AdaBoost allows designers to keep adding new weak classifiers until a predetermined error rate is small enough. The theory proves that as long as the classification ability of each weak classifier is better than the random guess, when the number of weak classifiers tends to infinity, the error rate of strong classifiers will tend to 0. In AdaBoost, every training sample is given a weight, indicating the probability that it will be selected by a component classifier into the training set. If a sample point has been accurately classified, its probability of being selected is reduced in constructing the next training set; conversely, if a sample point is not correctly classified, its weight is increased. With a few rounds of training, the AdaBoost can "focus on" the more difficult (more informative) samples and build strong classifiers for object detection. Hansen and Salamon proved that the integration method can effectively improve the generalization ability of the system. In practical application, since each independent classifier cannot guarantee the error is not relevant, the effect of classifier integration has a certain gap compared with the ideal value, but the role of improving generalization ability is still quite obvious.

6.4 Random forest

6.4.1 Principle description

Random forest, as its name implies, creates a forest in a random way. There are many decision trees in the forest. There is no correlation between each tree in a random forest. After getting the forest, when a new input sample comes in, let each decision tree in the forest make a judgment separately, observe which class the sample should belong to, and count which one is selected the most. Then, predict which class the sample belongs to, that is, select a mode as the final classification results (K., 1995).

6.4.2 Algorithm details

In establishing each decision tree, there are two things to be aware of: sampling and complete splitting. The first is two random sampling processes: input data of random forest to be carried out the row and column sampling. For line sampling, there is a way of putting it back, that is, there may be duplicate samples in the sampled sample set. Assuming that there are N input samples, there are also N samples for sampling. The input samples for each tree are not all samples, making them relatively less prone to over-fitting and then perform column sampling, randomly selecting m out of M classification features ($m \ll M$).

The data after sampling is completely split to establish a decision tree so that any leaf node in the decision tree either cannot continue to split or all the samples inside belong to the same class. An important step in many decision tree algorithms is pruning, but pruning is not used in the random forest. Since the two previous random sampling processes ensure randomness, over-fitting does not appear even if they are not pruned.

Every tree in a random forest obtained by this algorithm is weak, but all the trees combine to form a powerful classifier. This can be compared to a random forest: every decision tree is an expert skilled in a narrow field (since we randomly select m out of the M features for each decision tree to learn), so in a random forest, there are a number of experts who are proficient in different fields. For a new problem (new input data), they can look at it from a different angle, and analyze it, finally obtaining the result of voting by various experts.

6.4.3 Algorithms analysis

1. OOB error estimation
When we construct a single decision tree and extract only N samples with putting back randomly, so we can test the classification accuracy of this decision tree by

using non-decimated samples. These samples are about one-third of the samples. Thus, for each sample j, about one-third of the decision trees (SetT(j)) do not use it in the construction. We choose them to classify the sample. For all training samples j, we classify them by the forest in SetT(j) and observe whether their classification result is the same as the actual classification. The proportion of unequal samples is out-of-bag (OOB) error estimation. OOB error estimates proved to be unbiased.

2. Feature importance assessment

The importance of a feature is a concept difficult to define, because the importance of a variable may be related to its interaction with other variables. The random forest assesses the importance based on the increment of the prediction error when the OOB of the test feature is replaced and the OOB of the other features are not changed. When constructing random forests, each tree needs to be calculated one by one. In the classification algorithm, the random forest uses four methods to assess the importance of features.

Random forest improves the prediction accuracy without significantly increasing computation and can predict the effect of up to several thousand explanatory variables well, which is regarded as one of the best algorithms at present. It has many advantages.

(1) It performs well on the data set and is simple to achieve.
(2) In many current data sets, it has a great advantage over other algorithms.
(3) It can deal with data of high dimensions (a number of features) and does not have to perform feature selection.
(4) It can produce what features are important after training.
(5) Unbiased estimators are used for generalization errors when creating a random forest.
(6) Training speed is fast.
(7) During the training, the mutual influence of the features can be detected.
(8) It is easy to make a parallel method.

Summary

In real life, there are many types of small-set data. Collecting these kinds of information and using them have become a new research hotspot in data analysis. Machine learning is the appropriate tool to process such kinds of data. This chapter mainly introduces AdaBoost and random forest. AdaBoost is an iterative algorithm that refers to a particular method of training a boosted classifier. In the AdaBoost algorithm, different weak learners are trained on the same training set, and then they are assembled to construct a final strong classifier. AdaBoost is achieved by altering the data distribution. At each iteration of the training process, a weight is assigned to each sample in the training set equal to the current error on that sample. These weights can be used to inform the training of the weak learners. The outputs of the weak learners are combined into a weighted sum that represents the final output of the boosted strong classifier. AdaBoost is adaptive in the sense that subsequent weak learners are tweaked in favor of those instances misclassified by previous classifiers. Random forest is an ensemble learning method for machine learning because of its inherent characteristics and good classification performance. It operates by constructing a multitude of decision trees at training time and outputting the class that is the mode of the classes (classification) or mean predictions (regression) of the individual trees. A decision tree depends on a random vector, and all vectors in the random forest are independent identically distributed.

7 Compressed sensing

Introduction

Compressed sensing, also known as compressive sampling, sparse sampling, or compressive sensing, is a signal processing method that enables efficient signal acquisition and reconstruction and finds solutions to underdetermined linear systems (Donoho, 2006). This is based on the principle that it can exploit the sparsity of a signal and recover the signal from far smaller amounts of samples than required by the Shannon–Nyquist sampling theorem through optimization. There are two possible conditions for recovery: the first is sparsity, which ensures a signal to be sparse in some domain; and the second is incoherence, which guarantees sufficiency for sparse signals by using the isometric property.

The overall aim of applications of signal processing is to reconstruct a signal with a series of sampling measurements. This task is usually considered impossible because it is by no means possible to reconstruct a signal at a time when the signal is not being measured. Nevertheless, with some assumptions or prior knowledge, it has been shown that the signal can be fully reconstructed from a series of measurements using the acquiring process called sampling. Over time, engineers have improved their understanding of practical assumptions and methods of generalization. The initial breakthrough in signal processing came with the Nyquist–Shannon sampling theorem, which states that if the maximum frequency of an actual complex signal is less than half of the sampling rate or less than the sampling rate, the sinc interpolation can be used to completely reconstruct the signal. The main idea of the sampling theorem is that if the constraints on the frequency of the signal are specified in advance, a small number of samples are required to reconstruct the signal. Around 2004, Emmanuel Candès, Terence Tao, Justin Romberg, and David Donoho proved that with the prior knowledge of a signal's sparsity, the signal could be reconstructed using much fewer samples than that required by the sampling theorem. This proposed idea is the basis of compressed sensing.

7.1 Theory framework

The traditional signal acquisition, encoding, and decoding process is shown in Figure 7.1. The encoder carries on the sampling to the signal first, then transforms all the sampled values, encodes the amplitude and the position of the important coefficient therein, and finally stores or transmits the coded value; the signal decoding process is the inverse process of encoding, After the signal is decompressed, it is recovered after inverse transform. Using this traditional encoding and decoding method, the sampling rate of the signal should not be less than

https://doi.org/10.1515/9783110595567-008

Code

Signal

X —→ Sampling —→ Transform, compression decoding —→ Y

Decoding

Receiving data

—→ Y

Decompression, inverse transformation

Recovery signal —→ X

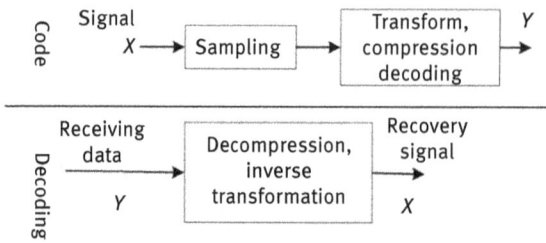

Figure 7.1: Block diagram of the traditional codec theory.

twice the bandwidth of the signal; this makes the hardware system face a great pressure of the sampling rate. In addition, during the process of compressing and encoding, a large number of transform-calculated small coefficients are discarded, resulting in data calculation and waste of memory resources.

The compressed sensing theory uses the same steps of sampling and compressing the signal as well as uses the sparsity of the signal to perform nonadaptive measurement coding at a rate well below the Nyquist sampling rate, as shown in Figure 7.2. The value is not the signal itself, but the projection from high to low dimension. From a mathematical point of view, each measurement is a combined function of each sample signal under the traditional theory that a measurement contains a small amount of all sample signals' information. The decoding process is not a simple inverse process of encoding but uses the existing reconstruction method in signal sparse decomposition to realize the accurate reconstruction of signal or approximate weight under certain error under the idea of inversion in a blind source separation structure. The number of measurements required for decoding is much smaller than the number of samples under the traditional theory.

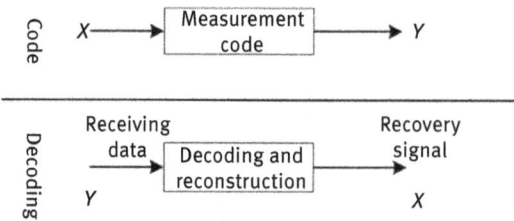

Code

X —→ Measurement code —→ Y

Decoding

Receiving data —→ Decoding and reconstruction —→ Recovery signal —→ X

Y

Figure 7.2: Block diagram of the compressed sensing theory codec.

Unlike the traditional Nyquist sampling theorem, the compressed sensing theory states that as long as the signal is compressible or sparse in a transform domain, the high-dimensional signal can be transformed using an observation matrix that is uncorrelated with the transform base. By projecting to a low-dimensional space and then reconstructing the original signal from these few projections with a high probability by solving an optimization problem prove that such a projection contains enough information to reconstruct the signal. In this theoretical framework, the sampling rate does not depend on the bandwidth of the signal but on the information in the signal structure and content.

7.2 Basic theory and core issues

7.2.1 Mathematical model

Suppose there is a signal $f\left(f \in R^{N*1}\right)$, with length N and base vector $\psi_i(i=1,2,\ldots,N)$; the signal can be transformed as follows:

$$f = \sum_{i=1}^{N} a_i\psi_i \text{ OR} f = \psi\alpha \tag{7.1}$$

Here, $\Psi=(\psi_1,\psi_2,\ldots,\psi_N) \in R^{N*N}$ is the orthogonal basis dictionary matrix, which meets the condition $\psi\psi^T = \psi^T\psi = I$; f is the representation of the signal in the time domain; and α is the representation of the signal in the ψ domain. Whether the signals have sparsity or approximate sparsity is the key issue of the theory of compressed sensing. If only K of α is nonzero (N ≥ K) (7.1) or exponentially decays after coming close to zero, the signal can be considered sparse. The sparse representation of a signal is an a priori condition of compressed sensing. In the known signal is compressed under the premise of the compressed sensing process can be divided into the following two steps.

(1) An M × N (M ≪ N) -dimensional measurement matrix that is not related to the transform basis is designed to observe the signal and an M-dimensional measurement vector is obtained.

(2) The signal is reconstructed from the M-dimensional measurement vector.

7.2.2 Signal sparse representation

Sparse mathematical definition: The transform coefficient vector of signal X under orthogonal basis Ψ is $\theta=\psi^TX$, if for $0<p<2$ and $R>0$, these coefficients satisfy

$$||\theta||_p \equiv \left(\sum_i |\theta_i|^p\right)^{1/p} \tag{7.2}$$

Then the coefficient vector Θ is sparse in a sense. If the potential of the support domain $\{i; \theta_i \neq 0\}$ of the transform coefficient $\theta_i = \langle X,\psi_i\rangle$ is less than or equal to K, the signal X can be said to be sparse. The best sparse domain is a premise condition for the application of the compressed sensing theory; only by selecting the appropriate base representation signal can signal sparsity be ensured, along with ensuring signal restoration accuracy.

Another hot spot in sparse representation research in recent years is the sparse decomposition of signals under redundant dictionaries (M., 2010). This is a completely new theory of signal representation: substituting the base function with an overcomplete library of redundant functions is called a redundant dictionary, and the elements of a dictionary are called atoms. The choice of the dictionary should be as good as possible to conform to the structure of the signal being approximated, which may be constructed without any restrictions. Find from the redundant dictionary K atoms with the best linear combination to represent a signal, called the signal sparse approximation or highly nonlinear approximation.

Currently, research on sparse representation of signals under redundant dictionaries focuses on two aspects: (1) how to construct a redundant dictionary suitable for a certain type of signals; and (2) how to design a fast and efficient sparse decomposition algorithm. These two issues have been studied and explored in depth. A series of theoretical proofs based on noncoherent dictionaries have been further improved.

7.2.3 Signal observation matrix

Linearly projecting the signal using an M × N (M ≪ N) measurement matrix ϕ that is not related to the transform matrix gives the linear measurement Y:

$$Y = \phi f \tag{7.3}$$

The measured value y is an M-dimensional vector, which reduces the measurement object from N-dimensional to M-dimensional. The observation process is nonadaptive, that is, choice of the measurement matrix does not depend on the signal f. Measurement matrix design requirements of the signal from f to y in the process and the K measured values will not undermine the original signal information, thus ensuring accurate signal reconstruction.

Because the signal f is sparsely expressed, eq. (7.3) can be expressed as

$$Y = \phi f = \psi \phi \alpha = \theta \alpha \tag{7.4}$$

Here, θ is an M × N matrix. In eq. (7.4), the number of equations is much smaller than the number of unknowns. The equation has no definite solution and cannot reconstruct the signal. However, since the signal is K-sparse, if Θ in eq. (7.4) satisfies the restricted isometry property (RIP), that is, for any K sparse signal f and the constant $\delta_k \in (0, 1)$, the matrix θ satisfies

$$1 - \delta_k \leq \frac{||\Theta f||_2^2}{||f||_2^2} \leq 1 + \delta_k \tag{7.5}$$

K coefficients can be accurately reconstructed from M measurements. The equivalent condition for RIP properties is that the measurement matrix ϕ and the sparse basis ψ are irrelevant. Presently, some of the measurement matrixes used for compressed sensing are Gaussian random matrix, binary random matrix (Bernoulli matrix), Fourier random matrix, Hadamard matrix, and uniform ball matrix.

Currently, the study of the observation matrix is an important aspect of the theory of compressed sensing. In this theory, the constraint on the observation matrix is relatively relaxed. Donoho provides three conditions that must be fulfilled by the observation matrix and points out that most of the stochastic matrixes that are uniformly distributed possess these three conditions and hence can be used as the observation matrix. Some examples are partial Fourier sets, partial Hadamard sets, uniformly distributed random projection sets, and so on, which are consistent with the conclusions drawn from the study of finite isometric properties. However, observations using the various observation matrices described previously can only guarantee the recovery of the signal with a high probability, but cannot accurately reconstruct the signal to 100%. Whether there is any real deterministic observation matrix for any stable reconstruction algorithm remains a problem to be studied.

7.2.4 Signal reconstruction algorithm

When the matrix Θ satisfies the RIP criterion, the compressed sensing theory can solve the sparse coefficient $\alpha = \psi^T x$ by solving the inverse problem in eq. (7.4) and then correctly recover the signal x with sparsity K from the measured projection value y. The most straightforward way to decode is to solve the optimization problem by the l_0-norm:

$$\min\|\alpha\|_{l_0} \quad s.t. \quad y = \phi\psi\alpha \tag{7.6}$$

Thus, an estimate of the sparse coefficient is obtained. Since eq. (7.6) is a nondeterministic polynomial (NP)-HARD problem, and its optimization problem is very similar to sparse decomposition of the signal, some scholars find a more effective solution from the theory of signal sparse decomposition. The literature (Extensions of Compressed Sensing, 2006) shows that the same solution can be obtained if the $l_1 l_1$-least-norm is equivalent to the l_0-least-norm under certain conditions. Then, eq. (7.6) can be transformed into an optimization problem under the l_1 minimum norm:

$$\min\|\alpha\|_{l_1} \quad s.t. \quad y = \phi\psi\alpha \tag{7.7}$$

In the optimization problem under the l_1 minimum norm, also known as the base track, the commonly used algorithms are the interior point method and the gradient projection method. The interior point method is usually slow but generates very accurate results, whereas the gradient projection method is fast but has less-accurate results than the interior point method. In the reconstruction of two-dimensional images, to make full use of the gradient structure of the image, it can be corrected to the total variation (TV) minimization method. Due to the slow algorithm under the l_1-min-norm, new fast greedy methods are gradually adopted, such as the matching pursuit method and the orthogonal matching pursuit method. In addition, the effective algorithms are the iterative threshold method and various improved algorithms.

7.3 Application and simulation

7.3.1 Application

The ability to efficiently gather information from compressible signals with a number of non-correlated measurements determines the effectiveness of compression-aware applications. Some examples are low-cost digital cameras and audio capture devices, energy-saving audio and image acquisition devices, astronomical observations, network transmission, military maps, and radar signal processing. The following summarizes the application of compression perception in several ways.

1. Data compressing
In some cases, the sparse matrix is unknown in the encoding or cannot be physically implemented in data compression. Since the measurement matrix does not need to be designed according to the structure of the code, the random measurement matrix can be considered as a general coding scheme that needs to be used only when decoding or rebuilding a signal. This versatility is particularly useful in the distributed coding of multi-signal devices such as sensor networks.

2. Channel coding
Compressed sensing sparsity, randomness, and convex optimization can be applied to the design of fast error correction code to prevent the wrong transmission.

3. Inverse problem
In other cases, the only way to obtain a signal is to use a specific mode of measurement system ϕ. However, given that the signal has a sparse transform base ψ and is uncorrelated with the measurement matrix ϕ, the signal can be effectively sensed. Such applications are mentioned in MR angiography, in which ϕ denotes a subset of Fourier transforms, and the resulting desired image signal is sparse in the time and wavelet domain.

7.3.2 Face recognition

1. Description of sparse representation and mathematical model

After the original signal is transformed to the discrete cosine transform (DCT) domain, only a very small number of elements are nonzero, while most of the elements are equal to zero or close to zero. This is the signal's sparsity. A sparse representation of a face refers to a face image that can be represented by a linear combination of face images owned by the same person in the face library. For other people's faces in the database, the linear combination coefficient is theoretically zero. Therefore, for the data in the face database to represent a person's face image, the coefficient vector should be sparse. That is, only the face image combination coefficient of the same person as that person is not zero, and all other coefficients are zero. Since face recognition based on sparse representation does not need training, the sparse representation of the dictionary can be directly used by the training of all the images; there are some improved learning algorithms for the dictionary. Since the sparse representation method is not sensitive to what features to use, only the original image data needs to be arranged into a large vector after simple processing and stored in the database. The sparse representation method can be abstracted by the following equation:

$$y = A \times X \tag{7.8}$$

Here, the represented sample y can be expressed by the coefficient vector X of the training sample space (or dictionary) A, and X is sparse, that is, most of its elements are zero or close to zero. Solving the sparse coefficient vector X is the process where y is sparsely represented. Since 0-norm represents the number of nonzero elements in the vector, the solution process can be simplified to

$$X_0 = \arg \min X_0 \quad s.t. y = A \times X \tag{7.9}$$

Solving the 0-norm X_0 minimization is an NP-hard problem. When x is sufficiently sparse, it can be replaced by a convex approximation of 1 norm optimization, that is, X_1 is

$$X_1 = \arg \min X_1 \quad s.t. y = A \times X \tag{7.10}$$

In the presence of noise and other non-ideal conditions, it can be solved by adding a relaxation error term $y = A \times x + e$. Equation (7.10) transforms to solve the following 1-norm problem:

$$X_1 = \arg \min X_1 \quad s.t. A \times X - y_2 < \rho \tag{7.11}$$

The problem of the whole sparse representation can be expressed by eq (7.11), that is, under the condition of $A \times X - y_2 \le e$, to find the solution X_1 of X when the norm-1 of X is the minimum. This solution algorithm is generally more time-consuming. Although many methods have been proposed, they still cannot meet the requirements of applications.

2. Face recognition via sparse representation
During face recognition, both the training set and the test samples are sparsely represented by a linear combination of the dictionaries. The test sample is cosine-distanced from all the samples in the training set, and the obtained largest value of the samples is the one that best matches the test sample of the people. The sparse representation of the face recognition process is shown in Figure 7.3.

$$\text{Sim}(x_i, x_j) = \cos\theta = \frac{x_i^T \times x_j}{x_{i2}x_{i2}} \tag{7.12}$$

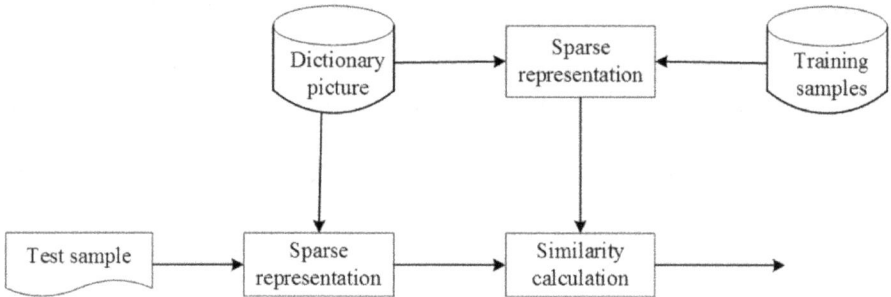

Figure 7.3: Sparse representation of the face recognition flowchart.

3. Improvement of the sparse representation method
(1) The whole sparse representation method is easily affected by the occlusion problem and the recognition rate is greatly reduced. Hence, we propose a sub-block representation based on compressed sensing. Here, the entire image is evenly divided into many equal areas, different regions are expressed separately, and finally they are voted together; the sum of all votes is considered the final recognition criteria, which helps address the problem of local occlusion. For a shaded face image, the sparse representation of the shaded part may not be accurate; however, only as a small part of the vote, it will not affect the overall voting result.

$$\text{Sim}(x_i, x_j) = \sum_{k=1}^{m} \frac{p_i^k \times p_j^k}{p_{i2}^k \, p_{j2}^k} \tag{7.13}$$

(2) Based on the research of nuclear thought in SVM, the feature of inseparable samples in low-dimensional space can be upgraded to high-dimensional and separable space. Recognition problem is challenging, due to lighting factors, occlusion problems, gestures, facial expressions so that the sample changes the characteristics of its space, identification or identification of errors equal to the original separable become inseparable, in view of this idea of using nuclear

SRC to further improve the performance of the above-mentioned method, that is, the sparse representation of nuclear processing block vote method, expressed in KB-SRC. Using the following kernel, n is a number greater than or equal to one:

$$k(x_i, x_j) = \left(x_i^T, x_j\right)^{\frac{1}{n}} \tag{7.14}$$

Summary

This chapter discusses three issues of the compressed sensing theory framework and technology. Compressed sensing theory is a signal processing technique that uses the sparsity of signal and transforms the original Nyquist sampling into a process where a signal can be exploited to recover it from far fewer samples through optimization. There are two possible conditions for the recovery: the first is sparsity, which ensures a signal is sparse in some domain; and the second is incoherence, which guarantees sufficiency for sparse signals by using the isometric property. It eliminates the process of sampling redundant data fast and then removing much unnecessary data, greatly reducing the cost of processing, saving, and transmission. Compressed sensing can be regarded as a novel sampling theory that fuses sampling and compressing in the same process.

8 Subspace learning

Introduction

Currently, feature extraction method based on principal component analysis (PCA) is a trending research topic. The method it represents is called the subspace learning method, and it is primarily used for feature extraction and has been successfully applied in the field of face recognition (Vasilescu, 2003) (I.T., 2002) (Zhang, 2018) (Linlin Yang, 2017).

In a pattern recognition system, feature extraction plays an important part. Feature extraction, as it is termed, essentially extracts the effective features from the input signal, one of the most important features being dimensionality reduction. In particular, feature extraction in face images extracts valid information from a given face image. Because, usually, the pictures are 64 × 64, the feature extraction is based on 83 points. The extracted geometric information from the points is identified, making the pattern recognition system relatively simple. There are different application feature extraction methods that are inconsistent; however, feature extraction based on PCA is a common method that is used in different application feature extraction methods.

8.1 Feature extraction based on PCA

PCA is a method of dimensionality reduction using linear mapping and removes the correlation of the data to maintain the variance of the original data.

We first review the linear mapping method. A linear mapping from a P-dimensional vector X to a one-dimensional vector F is formulated as

$$F = \sum_{i=1}^{p} u_i X_i = u_1 X_1 + u_2 X_2 + u_3 X_3 + \ldots + u_p X_p$$

This is equivalent to a weighted summation, with each set of weight coefficients being a principal component that has the same dimensions as the dimensions of the input data, such as $X = (1,1)^T, u = (1,0)^T$; hence, the linear mapping of two-dimensional vector X to one-dimensional space is

$$F = u^T X = 1 \times 1 + 1 \times 0 = 1$$

In the advanced algebra, the geometric meaning of F is expressed as the projection point of X in the projection direction u. That is, in the Cartesian coordinate system, the above-mentioned example can be expressed as a perpendicular to the abscissa.

PCA is based on linear mapping, and it is calculated as follows: X is a P-dimensional vector, and PCA is the process of transforming this P-dimensional original vector into a K-dimensional vector by linear mapping, where K ≤ P, which is

https://doi.org/10.1515/9783110595567-009

$$F_1 = u_{11}X_1 + u_{21}X_2 + u_{31}X_3 + \ldots + u_{p1}X_p$$
$$F_2 = u_{12}X_1 + u_{22}X_2 + u_{32}X_3 + \ldots + u_{p2}X_p$$

$$\ldots$$

$$F_k = u_{1k}X_1 + u_{2k}X_2 + u_{3k}X_3 + \ldots + u_{pk}X_p$$

For example, the two-dimensional vector $X = (1,1)^T$ becomes a one-dimensional new vector $F_1 = 1 \times 1 + 1 \times 0 = 1$ through the linear mapping $u_1 = (1,0)^T$.

In the meantime, to remove the correlation of the data, it is only necessary to make the respective principal components orthogonal, and at this moment, the space formed by the orthogonal basis is called subspace.

Regarding the example of PCA, a very well-known work was the 1947 study of the national economy by American statistician Stone. Using data from the United States for each of the years 1929–1938, he obtained 17 variables that reflect national income and expenditure, such as employer subsidies, consumption and means of production, net public expenditures, and net additions to inventories, dividends, interest trade balances, and so on. After PCA, the variance information of the original 17 variables was replaced by three new variables with an accuracy of 97.4%. Based on economic knowledge, Stone named the three new variables the total income F1, the rate of change of total income F2, and the trend of economic development or recession F3. This makes us maintain the variance in the low-dimensional space as much as possible in order to maintain the original spatial data variance. Sample variance is the average of the sum of squares of the differences between the data and the average sample in the data set. In addition, there is an approximate assumption in our discussion that fake data satisfy Gaussian distributions or approximately Gaussian distributions. PCA is based on the covariance matrix, and the reason for this needs to be pondered on.

Overall, the basic idea of feature extraction based on PCA is to try to reduce the dimensionality of data in high-dimensional space under the principle of minimizing the loss of data information. This is because identification systems are much easier in a low-dimensional space than in a high-dimensional space. In addition, it is required to be able to remove the correlation of the data so that efficient feature extraction can be carried on.

Let us consider two examples of PCA. The first is to reduce the dimensionality of 2D space points, as shown in Figure 8.1.

The larger the variance, the more dispersed is the data, and the more the distance information in the original space can be maintained. The variance formula is

$$\frac{1}{n}\sum_{l=1}^{n}(x_l - \bar{x})^T(x_l - \bar{x})$$

Geometrically, the projection direction is always the most distributive direction along the data distribution. To remove the correlation, the projection directions

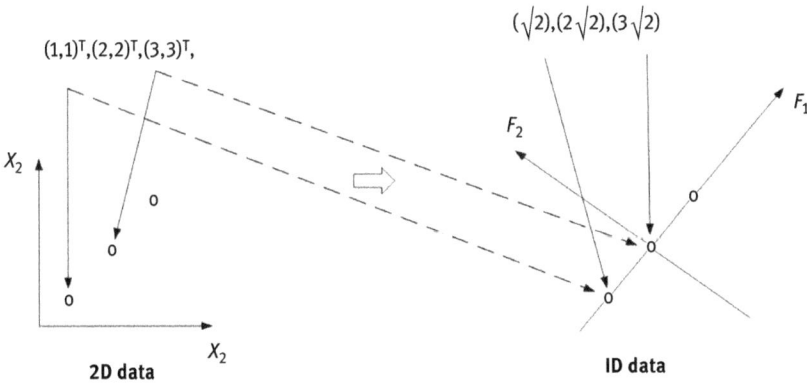

Figure 8.1: A PCA example.

should be orthogonal. In this case, the original data space category information is not lost, but the dimension is reduced by 50%.

To deepen our understanding, we discuss the geometric meanings of the principal components in a two-dimensional space. There are n samples, each of which has two dimensions, x1 and x2. In the two-dimensional plane defined by x1 and x2, the distribution of n samples is elliptical, as shown in Figure 8.2.

It can be seen from Figure 8.2 that the n samples have the largest dispersion along the F1 axis, which is the first principal component. To remove the correlation,

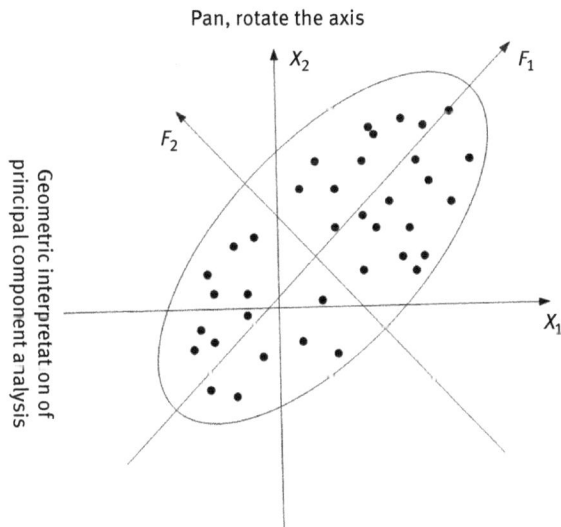

Figure 8.2: Two-dimensional spatial PCA.

the second principal component should be orthogonal to the first principal component. If you only consider one of F1 and F2, the information contained in the original data will be lost. However, depending on the accuracy of the system, you can
only select F1, as shown in Figure 8.3.

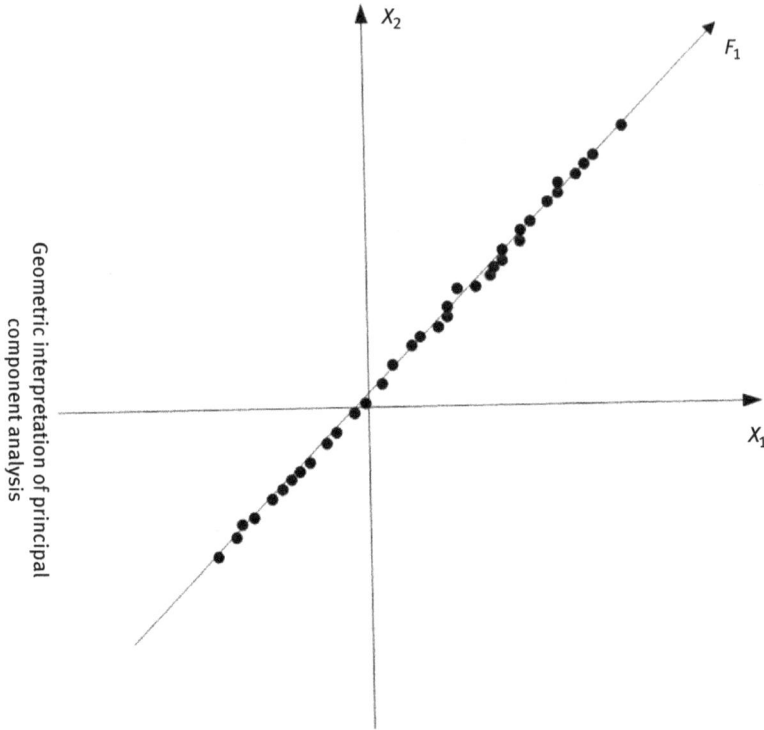

Figure 8.3: Geometric interpretation of two-dimensional PCA.

The practical problem always becomes a mathematical problem, which can then be
solved by the machine. We discuss the following mathematical model of PCA, we
arrange: X represents a variable; if X represents a vector, X_i represents the i-th component of the vector; if X represents a matrix, X_i represents the i-th component (column vector) of the matrix, and Xij represents the i-th component of the j-th sample.

8.2 Mathematical model

Assuming that we discuss the practical problem, X is a P-dimensional variable,
note as X_1, X_2, ..., X_p, PCA considers the problem of P variables into a linear combination of P variables, and these new components F_1, F_2, ..., F_k (k ≤ P), according

to the principle of retaining the main information to fully reflect the original varia-bles of the information, and independent of each other.

This process of reducing a variable from a multidimensional variable to a lower one is mathematically called dimensionality reduction. It is common operating for PCA to find a linear combination of vectors F_i.

$$F_1 = u_{11}X_1 + u_{21}X_2 + u_{31}X_3 + \ldots + u_{p1}X_p$$

$$F_2 = u_{12}X_1 + u_{22}X_2 + u_{32}X_3 + \ldots + u_{p2}X_p$$

$$\cdots$$

$$F_k = u_{1k}X_1 + u_{2k}X_2 + u_{3k}X_3 + \ldots + u_{pk}X_p$$

The following conditions need to be met.

(1) The square coefficient of the principal components' sum is one, as $u_{i1}^2 + u_{i2}^2 + \ldots + u_{ip}^2 = 1$.
(2) The principal components are independent of each other, with no overlap infor-mation, as $\text{Cov}(F_i, F_j) = 0$, $i \neq j$, $i, j = 1, 2, \ldots, p$.
(3) The variance of the principal components decreases in order, decreasing in im-portance as $\text{Var}(F_1) \geq \text{Var}(F_2) \ldots \geq \text{Var}(F_p)$.

8.3 Mathematical calculation of PCA

8.3.1 Conclusions of linear algebra

(1) If A is a p-order positive definite or semi-positive definite matrix, then the or-thogonal array U must be found:

$$U^T A U = \begin{bmatrix} \lambda_1 & \cdots & 0 \\ \vdots & \ddots & \vdots \\ 0 & \cdots & \lambda_p \end{bmatrix}_{p \times p}$$

Here, $\lambda, i = 1, 2, \ldots,$ p is the eigenvalues of A.

(2) If the eigenvalues of the above-mentioned matrix corresponding to the unit ei-genvectors are $u_1, \ldots, u_p$, we have

$$U = (u_1, \quad , u_p) - \begin{bmatrix} u_{11} & \cdots & u_{1p} \\ \vdots & \ddots & \vdots \\ u_{p1} & \cdots & u_{pp} \end{bmatrix}$$

The eigenvectors belonging to different eigenvalues corresponding to the real symmetric matrix A are orthogonal, so

$$U^T U = U U^T = I$$

8.3.2 Eigenvalue decomposition based on the covariance matrix

Because $F = u^T X, \bar{F} = \frac{1}{n}\sum_F F$, the following derivation process is established:

$$\text{Max:} \frac{1}{n-1}\sum_F (F-\bar{F})(F-\bar{F})^T = \frac{1}{n-1}\sum_x (u^T(x-\bar{x}))(u^T(x-\bar{x}))^T$$

$$\text{Max:} \frac{1}{n-1}\sum_x u^T(x-\bar{x})(x-\bar{x})^T u = u^T\left(\frac{1}{n-1}\sum_x (x-\bar{x})(x-\bar{x})^T\right)u$$

Constraint: $u^T u = 1$

Make $\frac{1}{n-1}(x-\bar{x})(x-\bar{x})^T = \Sigma x$

We introduce Lagrange multipliers and obtain the Lagrange function: $J(u) = u^T \sum_x u - \lambda(u^T u - 1)$, where λ is the Lagrange multiplier. Next, we find the partial derivative of u, and make the partial derivative equal to zero:

$$\frac{\partial J(u)}{\partial u} = 2\sum_x u - 2\lambda u = 0$$

$$\sum_x u = \lambda u \rightarrow u^T \sum_x u = \lambda$$

We consider Σx as the covariance matrix of X, so we assume

$$\sum x = \begin{bmatrix} \sigma_1^2 & \cdots & \sigma_{1p} \\ \vdots & \ddots & \vdots \\ \sigma_{p1} & \cdots & \sigma_p^2 \end{bmatrix}.$$

Because Σx is a symmetrical array, there is an orthogonal array U to obtain the following:

$$U^T \sum x U = \begin{bmatrix} \lambda_1 & \cdots & 0 \\ \vdots & \ddots & \vdots \\ 0 & \cdots & \lambda_p \end{bmatrix}$$

8.3.3 PCA

First, we agree: $\sum x = \left(\frac{1}{n-1}\sum_{i=1}^{n}(x-\bar{x})(x-\bar{x})^T\right)_{p\times p}$

$$X_i = \left(x_{1i}, x_{2i}, \ldots, x_{pi}\right)^T (i = 1, 2, \ldots, n)$$

First step: From the covariance matrix Σx of X, find its eigenvalue, that is, solve the equation $|\Sigma - \lambda I|$ and obtain the eigenvalue $\lambda_1 \geq \lambda_2 \geq \ldots \geq \lambda_p \geq 0$.

Second step: Find the corresponding feature vector $U_1, U_2, \ldots, U_p, U_i = \left(u_{1i}, u_{2i}, \ldots, u_{pi}\right)^T$.

Third step: Give the appropriate number of principal components. $F_i = U_i^T X$, $i = 1, 2, \ldots, k(k \leq p)$.

Fourth step: Calculate the scores of the k principal components selected. Centralize the raw data:

$$X_i = X_i - \bar{X} = \left(x_{1i} - \bar{x}_1, x_{2i} - \bar{x}_2, \ldots, x_{pi} - \bar{x}_p\right)^T$$

Substitute the expression of the first k principal components and then calculate for each k unit of the main score, and according to the size of the score queue.

Consider the three-point $(1,1)(2,2)(3,3)$ PCA, and find its eigenvectors and eigenvalues.

For the following known data sets:

$$\Omega_1 : (-5, -5)^T (-5, -4)^I (-4, -5)^I (-5, -6)^T (-6, -5)^T$$
$$\Omega_2 : (5,5)^T (5,4)^T (4,5)^T (5,6)^T (6,5)^T$$

compress the features from two dimensions to one dimension.

8.4 Property of PCA

1. Average value

$$E\left(U^T x\right) = U^T x$$

2. Variance is the sum value of all the eigenvalues

$$\lambda_1 + \lambda_2 + \ldots + \lambda_p = \sigma_1^2 + \sigma_2^2 + \ldots + \sigma_p^2$$

This shows that PCA decomposes the total variance of P-dimensional random variables into the sum of the variance of P uncorrelated random variables. The sum of the elements on the diagonal of the covariance matrix Σ is equal to the sum of the eigenvalues, that is, the variance.

3. Choosing the number of principal components

   (1) Contribution rate: Contribution rate is the proportion $\lambda_i / \Sigma_{i=1}^p \lambda_i$ of the variance of the i-th principal component in the total variance. The contribution rate

reflects the information of the original i eigenvectors and the extent of the ability to extract information.

(2) Cumulative contribution rate: how much the overall ability of the first k principal components, with the k principal components of the variance and the proportion of total variance $\Sigma_{i=1}^{k} \lambda_i / \Sigma_{i=1}^{p} \lambda_i$ to describe, known as the cumulative contribution rate.

One of the purposes of the PCA is to replace the original P-dimensional vector with as few principal components F1, F2,..., Fp(k ≤ p) as possible. In the end, how many principal components should be selected? In practice, the number of principal components depends on the amount of information that can reflect more than 95% of the original variable, that is, the number of principal components when the cumulative contribution rate achieves 95%.

Example: Assume that the covariance matrix of X1, X2, X3 is

$$\Sigma = \begin{pmatrix} 1 & -2 & 0 \\ -2 & 5 & 0 \\ 0 & 0 & 2 \end{pmatrix}$$

Because $|\Sigma - \lambda I| = 0$, the eigenvalue of the solution is $\lambda_1 = 5.83$, $\lambda_2 = 2.00$, $\lambda_3 = 0.17$.

Also, because $(\Sigma - \lambda I)U = 0$, that is $\Sigma U = \lambda U$, the solution is

$$U_1 = \begin{matrix} 0.383 \\ -0.924 \\ 0.000 \end{matrix}, \quad U_2 = \begin{matrix} 0 \\ 0 \\ 1 \end{matrix}, \quad U_3 = \begin{matrix} 0.924 \\ 0.383 \\ 0.000 \end{matrix}$$

Therefore, the contribution rate of the first principal component is 5.83/(5.83 + 2.00 + 0.17) = 72.875%. Although the contribution rate of the first principal component is not small, the first principal component does not contain information on the third original variable in this question; hence, two main components should be considered.

4. The correlation coefficient between the original variable and the principal component.

Because $F_j = u_{1j}x_1 + u_{2j}x_2 + \ldots + u_{pj}u_p$, and $j = 1, 2, \ldots, m$, $m \le p$, and $F = U^T X$, we can obtain

$$\begin{bmatrix} x_1 \\ x_2 \\ \vdots \\ x_p \end{bmatrix} = \begin{bmatrix} u_{11} & \cdots & u_{1p} \\ \vdots & \ddots & \vdots \\ u_{p1} & \cdots & u_{pp} \end{bmatrix} \begin{bmatrix} F_1 \\ F_2 \\ \vdots \\ F_p \end{bmatrix}$$

8.5 Face recognition based on PCA

Face recognition uses biometrics for recognizing purposes. A computer can confirm a person's identity by taking a person's face image or video as a research object. Recently, as a research topic that has both theoretical value and application value, face recognition has drawn increased attention from researchers, leading to the emergence of a variety of face recognition methods. PCA is one of them. Therefore, the key issue of face recognition is extracting features. One question that arises is how can the redundant information be removed and how can the face image be changed from a matrix into a vector. This can be answered using the method called PCA.

The calculation process of related research is shown in Figure 8.4.

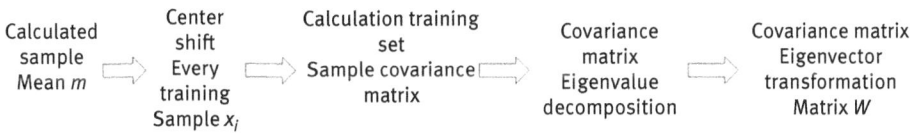

| Calculated sample Mean m | $\Rightarrow$ | Center shift Every training Sample x_i | $\Rightarrow$ | Calculation training set Sample covariance matrix | $\Rightarrow$ | Covariance matrix Eigenvalue decomposition | $\Rightarrow$ | Covariance matrix Eigenvector transformation Matrix W |

Figure 8.4: The calculation process.

The covariance matrix of the input training sample set is defined as

$$\sum_x = \frac{1}{n-1}\sum_{i-1}^{n}(x_i - \bar{x})(x_i - \bar{x})^T$$

where $\bar{x}$ is the average value of face sampling.

The PCA dimension reduction sorts the eigenvectors according to the eigenvalues corresponding to the eigenvectors. The eigenvectors corresponding to the largest K eigenvalues are selected to form the transformation matrix W, thus completing the projection from p dimensional space to k-dimensional space.

The 64 × 64 samples of the data set are shown in Figure 8.5, and the eight principal component features of the faces are shown in Figure 8.6.

Summary

PCA is a statistical method that converts a set of observations of possibly correlated variables into a set of values of linearly uncorrelated variables with principal components using an orthogonal transformation. PCA usually does well in dimension reduction of data. It is possible to sort new principal components in terms of importance, keep the most important components, and eliminate the others for dimension reduction and model compression. At the same time, it still keeps the key information of the original data. The advantage of PCA is it is parameter-free. There

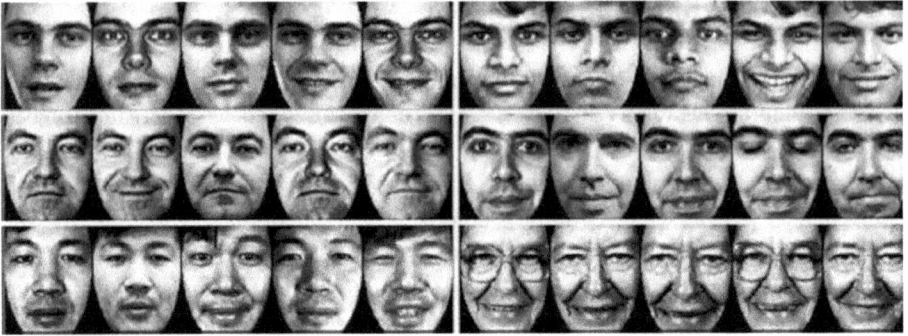

Figure 8.5: The data set.

Figure 8.6: Visualization of PCA components for the faces.

is no need to fix the parameters and use the prior model from experience; in other words, the final result of PCA only relies on the data instead of the user. From the other perspective, it may be recognized as a disadvantage. If users have certain prior knowledge of the observed data and masters some of their features, they would still face much difficulties in intervening in the process through parameterization; thus, the expected performance and high efficiency would not be achieved.

9 Deep learning and neural networks

Introduction

This chapter introduces the basic concepts of deep learning. Deep learning is an improved version of artificial neural network (ANN). Deep learning enhances the performance and supports the application of neural networks (Huang, 1996) (LeCun, 2015). First, this chapter introduces the basic model of a neural network, backpropagation (BP), and AutoEncoder algorithms for learning the model parameters in a supervised or unsupervised manner. They represent two common deep learning methods for vision problems.

9.1 Neural network

An ANN can be regarded as a directed graph with an artificial neuron as a node and a directed weighted arc. In this graph, the artificial neuron and the directed arc are the simulation of the biological neuron and the interaction of axon–synapse–dendrite, respectively. The weights of the directed arcs represent the strength index of the interaction between two connected neurons.

9.1.1 Forward neural network

The neurons in the feedforward process accept the previous level input and output it to the next level, without feedback; this can be represented by a directed acyclic graph. The nodes of the graph are divided into two types: the input node and the calculation unit. Each calculation unit can have any input, but only one output. The feedforward process is usually divided into different layers, and the input of the layer is connected only to the output of the layer. If the input node is the first layer, the network with a single layer computing unit is actually a two-layer network. The input and output nodes, open to end users, are directly affected by the environment, known as the visible layer, and the other intermediate layers are called hidden layers, as shown in Figure 9.1.

9.1.2 Perceptron network

The perceptron model is a kind of an extremely simple neural network model with learning ability, which was proposed by American scholar Rosenblatt for studying the brain's storage, learning, and cognition process.

https://doi.org/10.1515/9783110595567-010

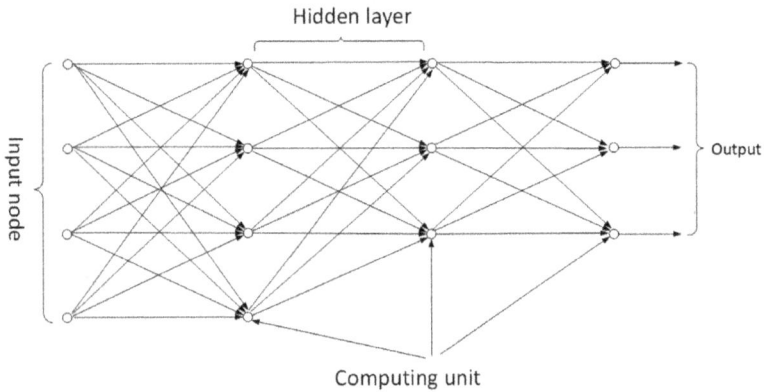

Figure 9.1: Schematic diagram of the structure of the feedforward neural network.

Perceptron is a double-layer neural network: an input layer and a computing layer. Perceptron can build discriminant ability through supervised learning, as shown in Figure 9.2.

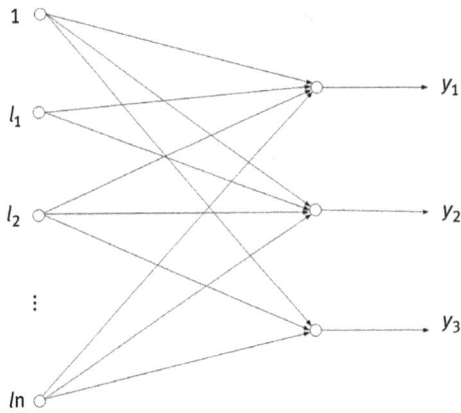

Figure 9.2: A perceptron model.

The goal of learning is to obtain a given output from a given input by changing the weight value. As a classifier, we can use the known pattern vectors and labels of the known classes as training sets. When the input belongs to the first class of feature vectors X, the corresponding first neuron in the output layer will be with the output $Y = 1$, whereas the output of other neurons is 0 (or −1). Setting the ideal output (m classes) to

$$Y = (y_1, y_2, \cdots, y_m)^T$$

The actual output is

$$\hat{Y} = (\hat{y}_1, \hat{y}_2, \cdots, \hat{y}_m)^T$$

To make the actual output close to the ideal output, we can input the vector X of the training set in turn and calculate the actual output $\hat{y}$, and make the following modification to the weight value represented by w:

$$w_{ij}(t+1) = w_{ij}(t) + \Delta w_{ij}(t), \tag{9.1}$$

where

$$\Delta w_{ij} = \eta(y_i - \hat{y}_j)x_i \tag{9.2}$$

The learning process of the perceptron is equivalent to the process of obtaining a linear discriminant function. Some characteristics of the perceptron are discussed here. ① The two-layer perceptron can only be used to solve the linear separable problem. ② The learning process can converge for any initial value for a linearly separated problem. In 1969, Minsky proved that the problem of "XOR" is a linearly non-separable problem. The definition of the XOR operation and the true value table of the corresponding logic operation are as follows.

x_1	x_2	y	x_1	x_2	y
0	0	0	1	0	1
0	1	1	1	1	0

$$y(x_1, x_2) = \begin{cases} 0, & x_1 = x_2 \\ 1, & other \end{cases}$$

If the "XOR" problem can be solved by a single-layer perceptron, the true value table of XOR can be known as w_1, w_2, and θ. The following must be satisfied:

$$\begin{cases} w_1 + w_2 - \theta < 0 \\ w_1 + 0 - \theta \geq 0 \\ 0 + 0 - \theta < 0 \\ 0 + w_2 - \theta \geq 0 \end{cases}$$

Obviously, they are unsolvable, which means that the single-layer perceptron is unable to solve the XOR problem. It is easy to analyze the problem with two inputs and two outputs. However, how much it is linearly separable or inseparable for a complex multi-input variable function remains unsolved Related research shows that the number of linear inseparable functions increases rapidly for increasing number of input variables, and even far exceeds the number of linear separable functions. That is, the number of problems that the single-layer perceptron cannot express is far more than the number of problems it can express. It explained that when Minsky revealed the fatal flaw in a single-layer perceptron, the development of ANNs entered a long and low ebb period.

9.1.3 Three-layer forward neural network

All kinds of logic gates can be realized using the nonlinear characteristics of artificial neurons. One example is the available threshold neuron gate, shown in Figure 9.3. Because the gate can determine any logic function which can be represented as disjunctive (conjunctive or combination of both), any logic function can be used as a three-layer (only a two-layer calculation unit) implementation of the feedforward network.

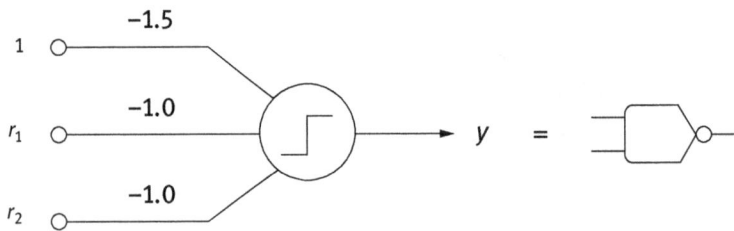

Figure 9.3: The "NAND" logic is realized by perceiving its model.

When the output function of neurons is a sigmoid function, the above-mentioned conclusions can be extended to continuous nonlinear functions. Under relaxed conditions, the three-level feedforward network can approximate any multivariate nonlinear function, breaking through the two-level feedforward network's linearly separable constraints. The feedforward network, which comprises three or more than three layers, is often called a multilayer perceptron (MLP).

9.1.4 BP algorithm

The three-level feedforward network is more applicable than the two-level feedforward network; however, the learning algorithm is more complex. The main difficulty lies in that the hidden layers are not directly open to the outside world, so they cannot be directly used to calculate the loss. To solve this problem, a BP algorithm is used. The main idea is to propagate the error of the output layer to the hidden-layer during the back forward process. The algorithm is divided into two stages. The first stage (forward propagation) input and output information from the input layer hidden the layer by layer-by-layer calculation of each unit value. The second stage (back propagation) according to the output error to calculate the forward layer hidden layer error of each unit, and the error correction layer before the right value.

The basic idea of the BP algorithm is to calculate the actual output O_k and error measure E_1according to the sample set (X_k, Y_k) one by one for the sample set $S = \{(X_1, Y_1), (X_2, Y_2), \cdots, (X_s, Y_s)\}$. The output layer weight matrix is adjusted by the error of the output layer, while the error estimates of all other layers are made through a back propagation process. The error in the output layer is back forward to the input. That is, one adjustment should be made to each $W^{(1)}, W^{(2)}, W^{(3)}, \ldots, W^{(n)}$, and this process should be repeated until $\sum E_p < \varepsilon$. In the BP algorithm, the gradient descent is used to update the weight, and also the output function is differentiable, e.g., the sigmoid function is usually used. Without losing its generality, we study the j computation unit at a certain level. The subscript i represents its previous layer i unit, k represents the k unit of the next level, O_j represents the output of the layer, and w_{ij} is the weight of the previous layer, as shown in Figure 9.4.

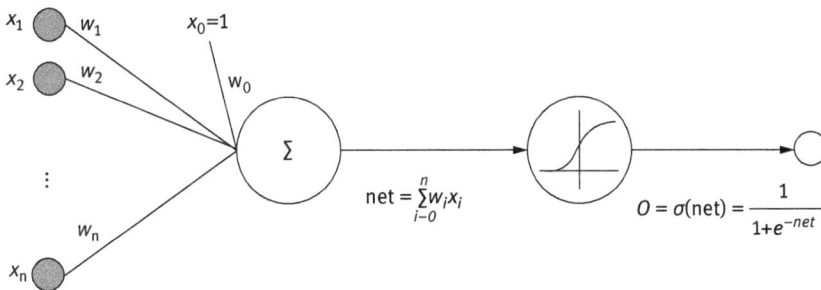

Figure 9.4: Neural network.

There are two parameters η and a' in the BP algorithm. The step size η has a great influence on the convergence. For example, η is tested between 0.1~0.3 and applies larger values for more complex problems. The coefficient term a affects the rate of convergence, and in many applications, its values can be selected between 0.9~1 (such as 0.95). There is a simple forward propagation network, as shown in Figure 9.5. When the BP algorithm is used to determine the weight of each connection, the calculation of δ is as follows.

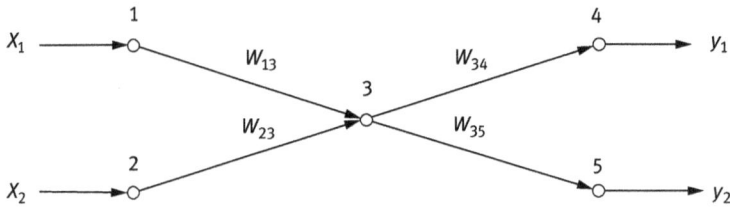

Figure 9.5: propagation simple network.

First, as we can see in Figure 9.5:

$$I_3 = W_{13}x_1 + W_{23}x_2 \quad O_3 = f(I_3)$$

$$I_4 = W_{34}O_3 \quad O_4 = y_1 = f(I_4)$$

$$I_5 = W_{35}O_3 \quad O_5 = y_2 = f(I_5)$$

$$e = \frac{1}{2}\left[\left(y'_2 - y_1\right)^2 + \left(y'_2 - y_2\right)^2\right]$$

The BP procedure is described as follows.

1. **Calculate** $\dfrac{\partial e}{\partial W}$

$$\frac{\partial e}{\partial W_{13}} = \frac{\partial e}{\partial I_3} \cdot \frac{\partial I_3}{\partial W_{13}} = \frac{\partial e}{\partial I_3} x_1 = \delta_3 x_1$$

$$\frac{\partial e}{\partial W_{23}} = \frac{\partial e}{\partial I_3} \cdot \frac{\partial I_3}{\partial W_{23}} = \frac{\partial e}{\partial I_3} x_2 = \delta_3 x_2$$

$$\frac{\partial e}{\partial W_{34}} = \frac{\partial e}{\partial I_4} \cdot \frac{\partial I_4}{\partial W_{34}} = \frac{\partial e}{\partial I_4} O_3 = \delta_3 O_3$$

$$\frac{\partial e}{\partial W_{35}} = \frac{\partial e}{\partial I_5} \cdot \frac{\partial I_5}{\partial W_{35}} = \frac{\partial e}{\partial I_5} O_3 = \delta_5 O_3$$

2. **Calculate** δ

$$\delta_4 = \frac{\partial e}{\partial I_4} = \left(y_1 - y'_1\right)f'(I_4)$$

$$\delta_5 = \frac{\partial e}{\partial I_5} = \left(y_2 - y'_2\right)f'(I_5)$$

$$\delta_3 = \left(\delta_4 W_{34} + \delta_5 W_{35}\right)f'(I_3)$$

That is, the calculation of δ_3 depends on δ_4 and δ_5 of the previous layers.

The number of nodes of output layer and of input layer of the three-layer feed-forward network are determined by the problem itself. For example, the node number of the input layer is the feature dimension, and the node number of the output is the category number. However, there is no effective method to determine the number of nodes in the hidden layer. In general, the more complex the problem is, the more hidden layer units are needed. Or, the same problem, the more the hidden layer units, the easier it is to converge. However, the excessive number of hidden layer units will increase the amount of calculation and will produce the "overfitting" problem, making the ability to discriminate the unseen samples even worse.

For the multi-class problem, the network output needs to divide the feature space into different class space (corresponding to different categories), and each hidden unit can be considered as a hyperplane. We know that the number of N hyperplanes can divide the D dimensional space into the number of regions:

$$M(N,D) = \sum_{i=0}^{D} N_i$$

When $N < D$, $M = 2^N$. There are P samples, and we do not know how many classes they should actually be divided into. For the sake of assurance, we can assume $M = P$. When $N < D$, the amount of hidden units is $N = \log_2 P$. The required number of hidden layer elements is mainly dependent on the complexity of the problem rather than on the number of samples.

The number of hidden layers is difficult to be determined but can be given first. Some hidden layer units could be pruned step by step to make the network more compact. The pruning principle will take into account the contribution of each hidden layer node or unit, for example, the size of the absolute value of each weight in the output layer or whether the input layer weight is similar to the other units. A more direct method is to delete a hidden layer node and then continue the learning process. If the network performance is obviously deteriorated, the original network is restored. The contribution of the hidden layer units is tested one by one, until all nodes less significant are deleted.

In principle, the BP algorithm can be used for any network. The multi-layer network can deal with any problem; however, for more complex problems, more layers of the network can achieve more better results. Unfortunately, when the BP algorithm is directly used for more than three-layer networks, it might fall into local minima and might not converge. It is necessary to use prior knowledge to reduce the search space or to find guiding principles to select the hidden layers.

The BP algorithm has a strong theoretical foundation, a rigorous derivation process, a clear concept, and good generalization; thus, it benefits training multi-layer networks. However, the convergence speed of a learning algorithm is still slow, how to choose the number of nodes in the network remains a pending problem.

From the mathematical point of view, the BP algorithm is even fast but suffering from the local minima problem.

9.2 Deep learning

9.2.1 Overview of deep learning

The concept of deep learning was put forward by Geoffrey Hinton et al. in 2006. It improved the traditional ANN algorithm and completed the recognition and classification of data by imitating how human brain processes signals. The "deep" in deep learning refers to the multilayer structure of the neural network. In traditional pattern recognition applications, the first step is preprocessing the data. Then, feature extraction is carried out on the preprocessed data. Based on these features, we train models with learning algorithms, e.g., SVM. Finally, the same kind of features are extracted from the test data as the input of the models, which are used for classification. In this process, feature extraction is a crucial step, and the selection of features directly affects the performance of the model for classification. In practical applications, designing the right features is a challenging task. Taking images as an example, there are many hand-crated features, such as scale-invariant feature transform (SIFT) and histogram of gradient (HOG). However, the deep learning methods can learn features from supervised or unsupervised aspects in raw data and then use the learned features as the inputs.

Although researchers had proposed ANNs in the 1980s, the application of ANNs have been limited for a long time. The learning ability of the shallow network is limited, during calculating the parameters of neural network models, the primary way is to randomize the weights of all the parameters of the initial network. Then, according to the principle of the minimum difference function over the training data, the gradient descent method is used to update the parameters. The method does not apply to the parameter training of the deep network, which often do not converge to the global optimum. Different parameter learning strategies are proposed for this problem. First, the network parameters are learned layer by layer and then optimization is carried out. Specifically, first by updating the model parameters, the output of the upper layer as the input of the next layer, the encoder layer produces the output, by adjusting the parameters to minimize the loss. . Finally, the BP algorithm can be used to adjust the parameters of the model, and all layers can be adjusted by a learning algorithm. The commonly used deep learning methods are stack AutoEncoder, convolutional neural networks (CNNs)restrict Boltzmann machine (RBM).

9.2.2 Auto-Encoder algorithm

The supervised learning of neural networks has been well conducted in the previous works. Unsupervised learning can also be used to train neural networks, which can learn a model in a self-supervised manner based on the BP algorithm, as shown in Figure 9.6.

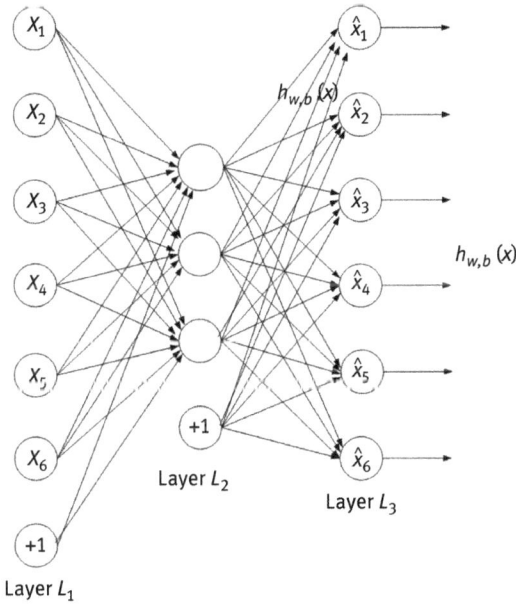

Figure 9.6: Auto-Encoder neural network.

The Auto-Encoder neural network aims to make the target mimic the input, that is, $h_{w,b}(x) \approx x$. The significance of the learning method lies in the data compression, which uses a number of neurons to reserve most of the information in the original data. It is similar to the data dimensionality reduction of principle component analysis (PCA). If the neuron is sparsely restricted in the middle hidden layer, some interesting phenomenon in the input data can be observed especially when the number of neurons is large. Sparsity can be interpreted, assuming that the activation is a sigmoid function. When the output of neurons is close to 1, it is activated otherwise inhibited. To achieve the full potential, an additional penalty factor can be added to the optimization objective:

$$KL(\rho \parallel \hat{\rho}_i) = \sum_{j=1}^{s_2} \rho \log \frac{\rho}{\hat{\rho}_i} + (1-\rho) \log \frac{1-\rho}{1-\hat{\rho}_i},$$

where ρ is a sparsity parameter. $\hat{\rho}_i = \frac{1}{m}\sum_{i=1}^{m}\left[a_j^{(2)}(x^{(i)})\right]$, where $a_j^{(2)}(x)$ indicates the activation of neuron J with x as the input. S is the number of neurons in the hidden layer. J represents each neuron in the hidden layer. The penalty factor is actually based on the concept of KL-entropy. Now, the overall cost function becomes:

$$J(W,b) = J(W,b) + \beta \sum_{j=1}^{s_2} KL(\rho \parallel \hat{\rho}_i),$$

where β controls the weight of the penalty factor, and $\hat{\rho}$ depends on W, b. The BP algorithm can be used to train parameters.

9.2.3 Auto-Encoder deep network

A three-layer neural network is introduced in this section, including the input layer, the hidden layer, and the output layer. It is still a very shallow network with only one hidden layer. This section discusses the deep neural network, that is, a neural network with multiple hidden layers.

The deep neural network can calculate more complex input features. This is because the activation function of each layer is a nonlinear function. Each hidden layer can exchange the output of the upper layer in a nonlinear way, and the deep neural network can learn more complex functional relations. The most important advantage of a deep network compared with a shallow network comes from that it can describe the signal in a more compact manner. However, the parameter training of the deep network does not use a simple gradient propagation algorithm. The reasons are as follows.

1) The above-mentioned methods require labeled data to complete the training. In some cases, getting enough labeled data is a high-cost task, and insufficient data will reduce the performance of the model.

2) This method can calculate more reasonable parameters for a less hierarchical network; however, for the deep network, it tends to converge to local extreme values instead of the global optimal solution.

3) The reason why the gradient descent method does not perform well in deep networks with randomly initialized weights is that, when using the BP algorithm to calculate derivatives, the magnitude of the gradient will decrease sharply with the increase of network depth, which is called gradient dispersion.

To solve these problems, the deep learning method first uses a meta-supervised learning method to learn the features without a large number of labeled data. After that, a layer-by-layer learning is adopted. We only train one layer of the network per epoch, with the parameters of each layer trained gradually; then, we use the BP algorithm to fine-tune the parameters of each layer. Compared with random initialization, the weights of each layer are well given, and it may converge to a better result.

1. **Stacked Auto-Encoder**

A simple is shown in Figure 9.7. We use Auto'-Encoder to train only one level of parameters. After training, we fix the parameters of the layer and output them as the input of the next layer to repeat this process. This allows one to derive the weight of each layer. This method is also known as the layer-by-layer greedy training. To obtain better results, after the above-mentioned training process is completed, we can adjust all the parameters of the layers to improve the result based on the BP algorithm. This process is called fine-tuning. For the classification tasks, usually we directly use the output of the encoder as the input of the softmax classifier, without the decoder in consideration.

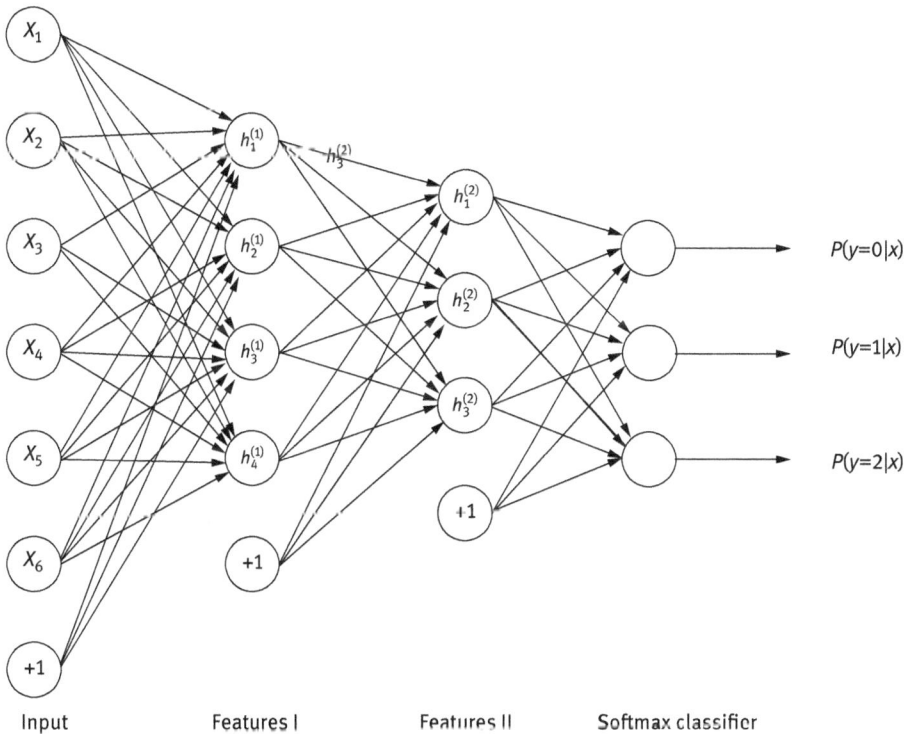

Figure 9.7: Auto Encoder .

9.2.4 Convolution neural network

Convolution neural network (CNN) is a kind of supervised deep model, which is especially suitable for processing two-dimensional data. Recently, new achievements and progress have been made in many applications, such as pedestrian detection, face recognition, signal processing. It is a deep neural network based on the convolution operation, which is also the first work to successfully train a multilayer network that is suitable for practical applications. The main difference between the CNN and the traditional ANN is the weight sharing and the non-full connection. Weight sharing can mitigate over-fitting, while establishing a non-fully-connected spatial relationship between different layers is used to reduce the number of training parameters which is also the basic idea behind CNNs. They uses back propagation algorithm to learn convolution kernels that can extract the intrinsic feature of the input data. The input data are convoluted with convolution kernels, followed by the pooling operation. With the deeper network architecture, the features gradually become abstract, and finally, the robustness to the translation, rotation and scaling invariance of the input data is obtained. Compared with the traditional neural network, CNNs execute the feature extraction and classification process simultaneously, benefiting real applications on the implementation.

The convolution layer of CNN is used to extract the local features of the input data, which can not only enhance the feature representation but also reduce the noise in the image. The sampling layer is used to downsample the input data to reduce the complexity leading to a certain invariance. In general, the convolution kernel of different sizes can be selected to extract the multi-scale features. The basic framework of CNN for image recognition is shown in Figure 9.8; it has two convolution layers, two down sampling layers, a full connection layer, and the output.

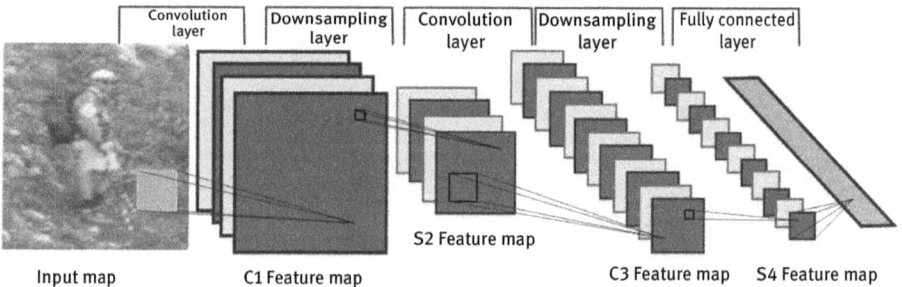

Figure 9.8: CNN framework.

The input image is convoluted with the kernels, and the feature (C1) is obtained through the activation function. The calculation formula of the convolution layer is as follows:

$$C_k^l = F\left(\sum_{n \in I_k} \omega_{nk} {}^* M_n^{l-1} + b_n^l\right)$$

Here, C_k^l is the k of the l-th feature map. I_k represents all the convoluted input images obtained from the k feature graph. ω_{nk} represents the learning parameters for the corresponding filter kernel. $*$ represents the convolution operation. M_n^{l-1} represents the n feature map of the $l-1$ layer. b_n^l is the first layer of the n input weighted bias of the corresponding image. $S(\cdot)$ is an incentive function of the coiling layer. It can be seen from the formula that the C1 feature map is composed of a plurality of input by accumulation and gain; however, for the same input graph, the convolution kernel parameter is the same, highlighting the significance of weight sharing. The initial value of the convolution kernel is not set randomly, but by unsupervised pretraining or according to a certain method, such as imitation of biological vision features or preprocessing with a Gabor filter. The convolution operation here is a two-dimensional discrete convolution operation for the image. The main step is to rotate the convolution kernel template 180 degrees first, then translate the center to the pixel points, multiply and accumulate the corresponding pixels, and finally obtain the convolution value of the pixel on the image. The lower sampling layer enhances the scaling invariance by reducing the spatial resolution of the network. The calculation formula is as follows:

$$S_k^l = F\left(\beta \sum_{n \in I_k} M_n^{l-1} + b_n^l\right)$$

Here, X represents a trained scalar parameter whose values vary with the lower sampling method. The common lower sampling method has the maximum lower sampling and the mean lower sampling. The former is more suitable for the extraction of image textures, and the latter can preserve the image background well. For example, the mean sampling $\beta = 1/m$ is used to indicate the downsampling of the $m \times m$ pixel block. Thus, for the output image, each dimension is $1/m$ of the original graph, each output graph has a weighted bias b_n^l, and then the output result is input to a nonlinear function (such as a sigmoid function). The outputs of CNN are generally adopted in the linear fully connected layer. At present, the most commonly used classification methods are logic regression and Softmax classification. The training process of CNN is similar to that of the traditional ANN. The BP algorithm is used with two important stages of forwarding propagation and BP. Given N training samples and C classes, the error functions are defined as follows:

$$E^N = \frac{1}{2}\sum_{n=1}^{N}\sum_{k=1}^{C}\left(y_k^n - t_k^n\right)^2.$$

Here, y_k^n is the network output of the n-th dimension of the k-th sample. t_k^n is the corresponding expectation. The error function E^N is the accumulation of the two losses. The parameter training process is like the stochastic gradient descent (SGD) algorithm. CNN encounters many problems in practical applications, such as pretraining the parameters of the network, convergence conditions, and incomplete connection. All of them need to be considered and optimized in practical applications.

This book introduces a Boosting-like CNN algorithm proposed by us, details of which can be found in the relevant papers previously published by us. Assuming a penalty weight α is added to the input sample, there is a linear relationship between the input u^l of the l-th layer and the output x^{l-1} of the previous layer:

$$u^l = \alpha \omega^l x^{l-1} + b^l, x^l = f(u^l)$$

Here, ω^l is the weight of the output layer. b^l is offset. The training process is constantly adjusted. x^{l-1} is the input of the upper layer, that is, the input of the layer. f is the excitation function of the output layer, which is generally sigmoid or hyperbolic tangent function. The sensitivity of the output layer is obtained by the derivation

$$\delta^l = f'(u^l) * (y^n - t^n)$$

The derivative of the error E to the weight value W^l is as follows:

$$\frac{\partial E}{\partial \omega^l} = \delta' \frac{\partial u}{\partial \omega} = x^{l-1} f'(u^l) * (y^n - t^n)\alpha$$

Finally, the δ updating rule is used for each neuron to update the weights as follows:

$$\omega^{l+1} = \omega' - \eta x^{l-1} f'(u^l) * (y^n - l^n)\alpha$$

Here, η is the learning rate, so we can achieve the update method of the weight value ω. The CNN itself can be considered as a series feature extractors, with each layer as a feature extractor, where the features are extracted from a low level to a high level, and the feature extraction results interact with each other. The classification results of a feature extractor and a former relationship are restricted by a layer-after-layer feedback. Assuming that CNN has n stages, the n classifier can be obtained by training the classifier in n different stages. Therefore, the Boosting algorithm, which constantly adjusts the distribution of sample weights during training, helps provide better feedback information for different network hierarchies so as to improve the network performance and make the network more stable. We allocate the feedback weight of the correct and error discriminant samples based on the output y^n and feedback from the last layer of the network to the beginning of the network.

$$od_{t+1} = \begin{cases} |o_t - y_t|\alpha_r, |o_t - y_t| < 0.5 \\ |o_t - Y_t|\alpha_\omega, |o_t - y_t| \geq 0.5 \end{cases}$$

Here, o_t is the actual detection value of the network. y_t is the label value of the sample. od_i is the sensitivity δ of the output layer. α_r and α_w are the penalty coefficients of the wrong classification samples and the correct classification samples, respectively. Since the final classifier in this paper uses a logistic regression function, the range of the output value is $(0, 1)$. Therefore, if $|o_t - y_t| < 0.5$, the classification is correct, and vice versa. When the sample is classified, the penalty weight is increased; on the other hand, the weight of the sample is reduced when the sample is correctly classified. This idea is similar to Boost – to train the neural network by constantly updating the weight of the sample, which can avoid the over-fitting of the network and then make the performance stable. The solution process of α_r and α_w is very critical. A parameter solution process is an adaptive selection of the parameter method. According to the discriminant condition of each sample, its contribution is determined. So, after each iteration, there will be a redistribution of sample weights to D_i. The error function of CNN is the optimization goal. Thus, the idea of Boosting is integrated into the convolution training, which improves not only the performance of the system but also the stability. The specific operation steps are as follows. First, the weight distribution of the initialization sample is

$$W_1 = (w_{11}, \cdots w_{1i}, \cdots, w_{1N}), \; w_{1i} = \frac{1}{N}, \; i = 1, 2, \cdots, N$$

For the number of training iterations $m = 1, 2, \cdots, N$, the classifier $G_m(x)$ uses a sample with weighted distribution D_m as its training data. The $G_m(x)$ classification error rate $e_m = P(F_m(x_i) \neq y_i) = \sum_{i=1}^{N} w_{mi} I(F_m(x_i) \neq y_i)$

Here, $I(x, y)$ is an indicator function, and w_{mi} is the weight of the i times, updating the weight distribution of the training data set.

$$W_{m+2} = (w_{m+1, 1}, \cdots w_{m+1, i}, \cdots w_{m+1, N}) w_{m+1, i}$$
$$= \frac{w_{mi}}{N_m} \exp(-\beta_m y_i F_m(x_i))$$

Among them, β_m is a coefficient that characterizes the classification of classifiers. N_m is a normalization factor, $N_m = \sum_{i=1}^{N} w_{mi} \exp(-\beta_m y_i F_m(x_i))$. In the training process, we use the weight distribution of the sample D_m as the parameters in the new CNN. It is worth noting that the Boosting-like algorithm has an obvious effect on the convergence stability.

9.3 Applications of deep learning

9.3.1 Binarized convolutional networks for classification

Deep convolutional neural networks (DCNNs) have attracted much attention due to their capability of learning powerful feature representations directly from raw pixels, thereby facilitating many computer vision tasks. However, its success has come up with a significant amount of model parameters and training cost. For instance, the sizes of most DCNN models for vision applications are easily beyond hundreds of megabytes, which restricts their practical usage in most embedded platforms. To this end, compressing CNNs has become a hot research topic, in which the binarization-based compression schemes have received an ever-increasing focus due to their high compression rate. Wang proposed a new scheme, termed modulated convolutional networks (MCNs), toward highly accurate binarized compression of CNNs (Wang, 2018). In principle, MCNs decompose the full CNNs (conventional CNNs, VGG, AlexNet, ResNets, or Wide-ResNets) into a compact set of binarized filters. In MCNs, binarized filters are optimally calculated based on a projection function and a new learning algorithm during the backpropagation. A new loss function, which jointly considers the filter loss, center loss, and softmax loss, is used to train MCNs in an end-to-end framework. The modulation filters are introduced to recover unbinarized filters from binarized filters, which leads to a new architecture to calculate the network model. MCNs can reduce the required storage space of convolutional filters by a factor of 32, in contrast to the full-precision model, while achieving much better performance than the state-of-the-art binarized models. Astonishingly, MCNs achieve a comparable performance to the full-precision ResNets and Wide-ResNets.

For further resource-constrained environments, Gu introduced projection convolutional neural networks (PCNNs) with a discrete backpropagation via projection to improve the performance of binarized neural networks (Gu, 2019). In PCNNs, the projection function is exploited for the first time to efficiently solve the discrete backpropagation problem, which leads to a new highly compressed CNN. By exploiting multiple projections, PCNNs learn a set of diverse quantized kernels that compress the full-precision kernels in a more efficient way than those proposed previously and achieve the best classification performance compared with other state-of-the-art binarized neural networks on the ImageNet and CIFAR data sets.

9.3.2 Time-series recognition

Due to the complex spatio-temporal variations of data, time-series recognition remains a challenging problem for the present deep networks. Xie proposed end-to-end hierarchical residual stochastic (HRS) networks to effectively and efficiently describe spatio-temporal variations. Specially, stochastic kernelized filters are designed in HSR based on a hierarchical framework with a new correlation residual block to align the spatio-temporal features of a sequence. HSR further encodes complex sequence patterns with a stochastic convolution residual block, which employs the stochastic kernelized filter and a dropout strategy to reconfigure the convolution filters for large-scale computing in deep networks. Experiments on large-scale time-series recognition data sets, namely, NTU RGB+D, SYSU-3D, UT-Kinect, and Radar Behavior, show that HRS networks significantly boost the performance of time-series recognition and improve the state-of-the-art of skeleton, action, and radar behavior recognition performance (Xie, 2019).

Considering the efficiency of recurrent neural networks for time-series representation, recent advances in complex user interaction, pose recognition, and skeleton recognition are developed by integrating bidirectional long-short term memory (BiLSTM) and bidirectional gated recurrent unit (BiGRU) with the Fisher criterion (Li, 2018). These discriminative models can classify the user's gesture effectively by analyzing the corresponding acceleration and angular velocity data of hand motion. On the other hand, Xie brought three powerful tools, including recurrent neural networks, convolutional neural networks, and attention mechanism, under the same umbrella and developed an efficient framework to investigate a new hypothesis of "memory attention + convolution network" for skeleton-based action recognition (Chunyu Xie, 2018). A temporal-then-spatial recalibration scheme in memory attention networks (MANs) is proposed to eliminate complex variations. MANs have two vital modules in the proposed architecture: temporal attention recalibration module (TARM) and spatio-temporal convolution module (STCM). Specifically, the TARM originates from residual learning and employs a novel attention learning network to recalibrate the temporal attention of frames in a sequence; then, the STCM uses the attention calibrated skeleton joint sequences as images and leverages the CNNs to further model the spatial and temporal information of skeleton data. These two modules (TARM and STCM) form a single network architecture that can be trained in an end-to-end manner together.

Summary

Deep learning is known as deep structured learning or hierarchical learning that automatically learns and classifies the low-level or high-level features. For example, for machine vision, deep learning methods learn the low-level representation from the original image, such as edge information. Then, they obtain the high-level representation with the linear or nonlinear combination of these low-level representations. Deep learning is recognized as a better representational learning method in terms of cascading multiple levels and plenty of parameters. It can deal with large-scale data, so it is suitable for image and voice problems without obvious features and can achieve good performance. In addition, the deep learning method integrates feature extraction and classification into an end-to-end framework, which learns features from data and reduces the workload of hand-crafted features. With better performance and more convenience, deep learning has been a hot topic in the field of machine learning.

10 Reinforcement learning

Introduction

Reinforcement learning is an important machine learning method that has many applications in the fields of intelligent control robot, analysis, prediction, and so on. There is no mention of reinforcement learning in the traditional categories of machine learning; however, in the study of connectionist learning, the algorithms are divided into three categories, namely, unsupervised learning, supervised learning, and reinforcement learning (Busoniu, 2017) (Sutton, 2018).

10.1 Overview of reinforcement learning

Reinforcement learning refers to learning from the environmental state to action mapping to maximize the cumulative reward value of the system's behavior from the environment (Richard S Sutton, 1998). In reinforcement learning, we design the agent's action selection algorithms to convert the external environment and state to the way of maximizing the rewards. Although the agent is not directly told what to do or what action to take, it can find and select the action that gets the most reward. The agent's action not only affects the immediate reward but also impacts the subsequent action and the final reward. The searching of trial and error and delay reinforcement are the two most important characteristics of reinforcement learning.

Reinforcement learning is the strategy that obtains the reward from the environment, runs the action according to the reward, and converges the data constantly for the optimization. As we understand it, the framework of learning strategy on reinforcement learning is shown in Figure 10.1.

Reinforcement learning is the interaction between the agent and the environment, which selects and performs different actions considering the value of the rewards, and finally maximizes the obtained rewards. In the interaction model, the different policies are arranged based on the different rewards. All elements of the model are shown in Figure 10.2.

As defined by Richard S. Sutton, the elements of reinforcement learning are a policy, a reward function, a value function, and a model of the environment.

https://doi.org/10.1515/9783110595567-011

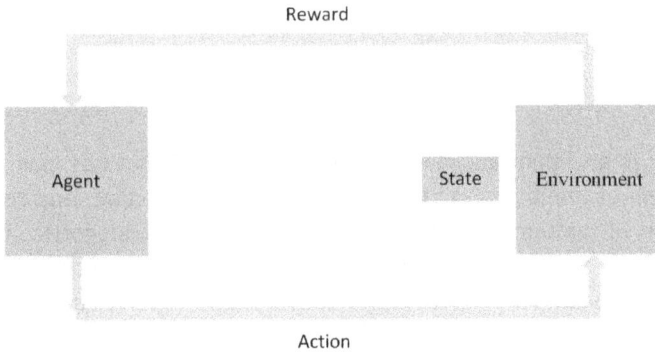

Figure 10.1: Framework of learning strategy.

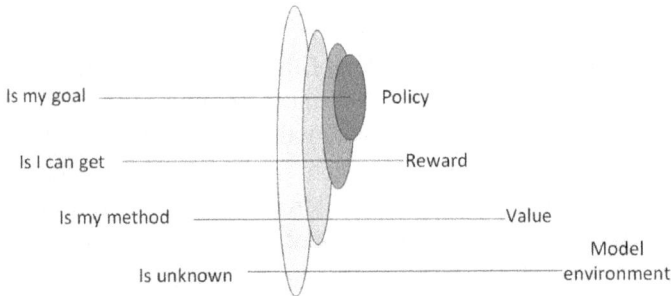

Figure 10.2: Elements of reinforcement learning.

10.2 Process of reinforcement learning

10.2.1 Markov property

The Markov process, named by Russian mathematician Andre Markov, is a discrete-time stochastic process with Markov property in mathematics. In this process, in the case of a given current knowledge or information, the conditional probability distribution of future states of the process depends only on the present states, not on the sequence of past states (events that preceded).

For reinforcement learning, the following probability is equivalent to:

$$\Pr\{s_{t+1}=s',r_{t+1}=r|s_t,a_t,r_t,s_{t-1},a_{t-1},\ldots,r_1,s_0,a_0\}$$

$$\Pr\{s_{t+1}=s',r_{t+1}=r|s_t,a_t\}$$

The decision-making process based on the return of reward is also the Markov decision process (MDP).

10.2.2 Reward

The reward is given by the environment, which differs according to the selection of
actions, and we define the expected return of reward as R_t:

$$R_t = r_{t+1} + r_{t+2} + r_{t+3} + \ldots + r_T$$

However, the decision process has less impact on time t with progress of time;
hence, the process is redefined with the discount rate (DR) as follows:

$$R_t = r_{t+1} + \gamma r_{t+2} + \gamma^2 r_{t+3} + \ldots + r_T$$

10.2.3 Value function

The determination of the value function is derived from the simplified decision-
making process as shown in Figure 10.3.

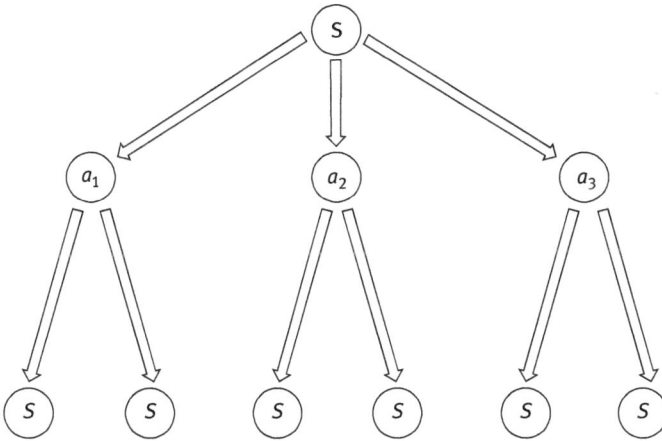

Figure 10.3: Simple example of the decision-making process.

In Figure 10.3, after each state S takes the actions a_1, a_2, a_3, each transition probability of S is $P^a_{ss'}$. The calculation of value is formulated as follows:

$$V^\pi(s) = \sum_a \pi(s,a) \sum_{s'} P^a_{ss'} + \gamma V^\pi(s')$$

10.2.4 Dynamic programming

Dynamic programming is an important progress in reinforcement learning. It simu-
lates all the future dynamic processes and selects the final optimal reward to back
up all values at each state under the condition that all states and actions are known.
The implementation is shown in Figure 10.4.

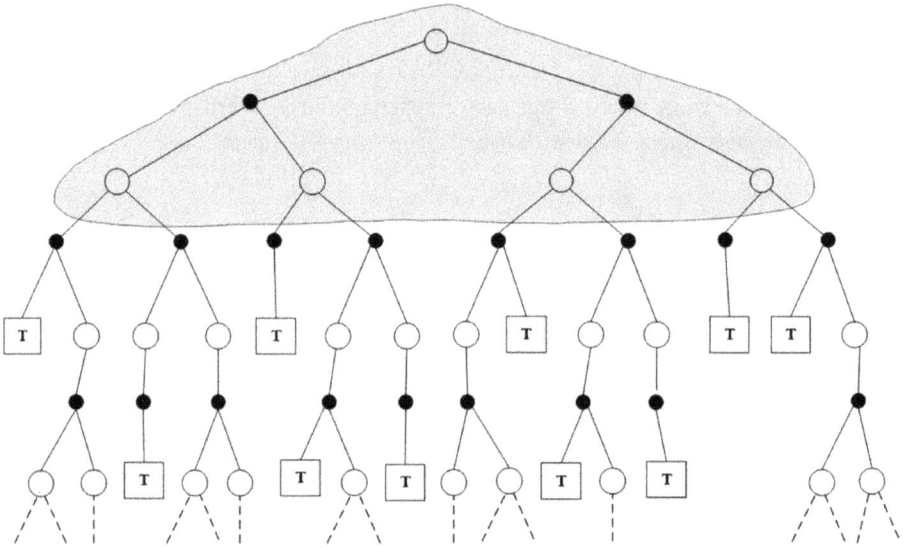

Figure 10.4: Dynamic programming.

Therefore, the dynamic programming method is optimized by iteratively generating
solutions to bigger subproblems by using the solutions to small subproblems; however, it needs to pay the cost of modeling and memorizing all states and actions.
For the process with many states and actions, the dynamic programming method is
difficult to model and replaced by a new kind of Monte Carlo (MC) method.

10.2.5 MC method

Usually, MC methods can be roughly divided into two categories: one is that the
problem has natural randomness, which can be directly simulated by the computing
power of the computer, for instance, study on the transmission process of neutrons
in the reactor in the field of nuclear physics. The interaction between neutrons and
nuclei is restricted by the quantum mechanism. Although the probability of their in-
teractions can be obtained, the accurate fission position of neutron and nuclear nuclei as well as the speed and direction of the new neutrons generated by fission will

not be obtained. Scientists randomly sample from a probability distribution the fission position, velocity, and direction. After simulating the great number of possible actions of neutrons in this manner, the range of neutron transmissions can be estimated by statistics as the basis for designing the reactor.

The other is that the problem can be transformed into the characteristics in the random distribution, such as the probability of occurrence of random events, or the expected value of random variables. The probability of random events is estimated by the random sampling method, or the numeric characteristics of random variables are estimated by the characteristics of sampled data, which is used as a solution to the problem. This method is mostly used to solve the complex multidimensional integral problem. To calculate the area of an irregular figure, the irregular degree of the graph should be proportional to the complexity of the analytical calculation. The MC method is based on this idea: suppose you have a bag of beans, sprinkle beans evenly on this graph, and then count the number of beans in the area in this figure. When your bean is smaller, the more it is sprinkled, the more accurate the result will be. With the help of computer programs, a large number of evenly distributed coordinate points can be generated, and then the number of points in the graph can be counted, and the figure area can be calculated by the proportion of total points they occupy and the area of the coordinates generated.

When applying MC methods in reinforcement learning, the used sampling method is resampling, which estimates the model based on sampling and the instance method instead of modeling all processes directly. It samples the data by maximizing the value state in a greedy or in other ways, and there are on-policy and off-policy techniques to implement it. You will ignore the difference for this book. The MC diagram is shown in Figure 10.5. The value state is updated as follows:

$$V(s_t) \leftarrow E_\pi \{ r_{t+1} + \gamma V(s_{t+1}) \}$$

10.2.6 Temporal difference learning

The temporal difference (TD) learning algorithm is a combination of MC and dynamic programming (DP). Similar to MC, it directly learns from the original experience without the dynamic information in the external environment. According to the different updating equations, different TD learning algorithms can be obtained. Among them, the tabular TD(0) method, as one of the simplest TD methods, estimates the value function by

$$V(s_t) \leftarrow V(s_t) + \alpha [r_{t+1} + \gamma V(s_{t+1}) - V(s_t)] \qquad (10.1)$$

In eq. (10.1), α is called the learning rate, and γ is the DR. In fact, in this case, the goal of TD is $r_{t+1} + \gamma V(s_{t+1})$, and $V(s_t)$ is updated on the basis of $V(s_{t+1})$. As the

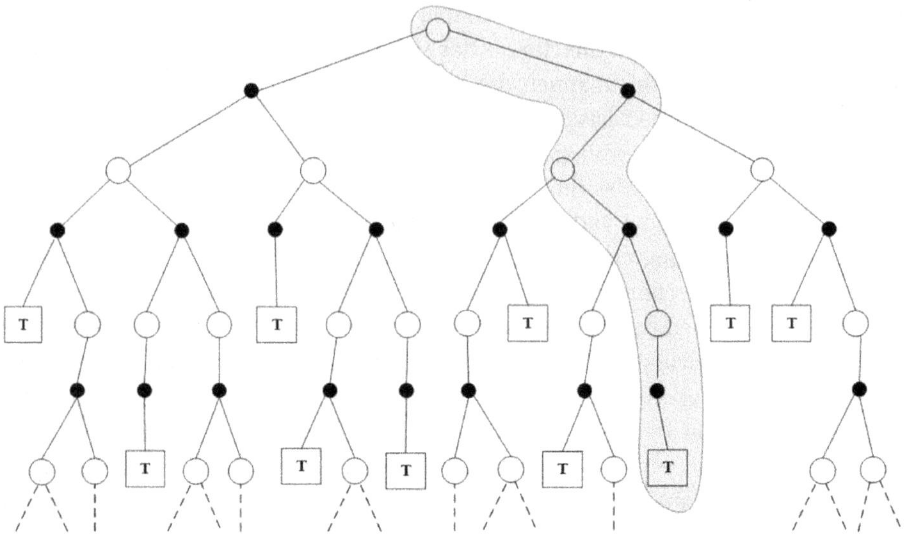

Figure 10.5: Visualization of the MC strategy.

dynamic programming method calculates a state value function based on the subsequent states, this method is also a step-by-step method.

In the strategy of initialization for TD(0), it takes the sample rewards as the value like MC methods and performs updates at the next time step by using the value function of the next states and the immediate rewards of $r_{t+1} + \gamma V(s_{t+1})$, which is different from the updates of the value function after a period. For TD(0), the simplest one in the TD algorithm, the backtrack algorithm, is shown in Figure 10.6.

The TD(0) algorithm is shown in Figure 10.7.

10.2.7 Q-learning

Q-learning is a model-independent reinforcement learning algorithm proposed by Watkins, which mainly learns a policy for the problem in MDP. After Watkins proposed and proved the convergence in 1989, the algorithm has drawn widespread attention.

Temporal-difference updated Q-learning is simply developed from the theory of dynamic programming and is a method of delayed reinforcement learning. In Q-learning, the policy and value functions are represented by a two-dimensional query table indexed by the state action pair. For each state x and action a:

$$Q^*(x, a) = R(x, a) + \gamma \sum_y P_{xy}(a) V^*(y), \tag{10.2}$$

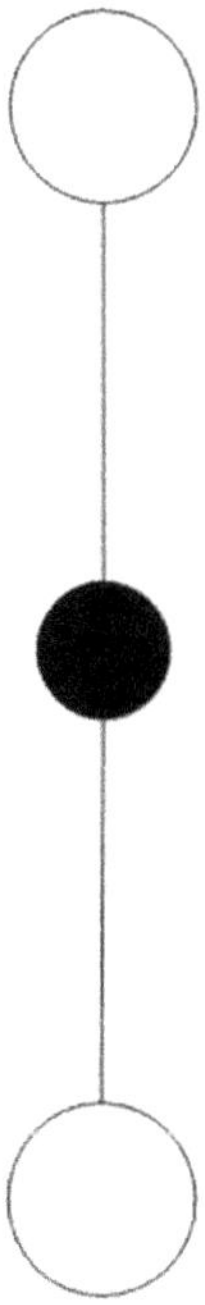

Figure 10.6: TD(0) backtrack algorithm.

Initial V(s) arbitrarily, π to the policy to be evaluated
Repeat(for each episode):
Initialize s
Repeat(for each step of episode):
    a ← action given by π for s
    Take action a; observe reward r, and next state'
    $V(s) \leftarrow V(s) + a[r + \gamma V(s') - V(s)]$
    s ← s'
    until s is terminal

Figure 10.7: TD(0) algorithm.

where $R(x, a) = E\{r_0 | x_0 = x, a_0 = a\}$. $P_{xy}(a)$ is the probability that the state moves from x to y when performing the action a. Equation (10.2) should satisfy the formula $V^*(x) = \max_a Q^*(x, a)$.

The estimated value $(\hat{Q}^*)$ of the Q function is evaluated by the Q-learning algorithm. The algorithm updates the $\hat{Q}^*$ value (often simply called the Q value) according to the performed action and the obtained reward value. The update of the $\hat{Q}^*$ value is based on the prediction deviation or the TD error of the Sutton. The difference between the discount value of the next state and the Q value of the current state action is

$$r + \gamma \hat{V}^*(y) - \hat{Q}^*(x, a) \tag{10.3}$$

Among them, r is the reward value. y is the next state that is migrated to the state x execution action a, $V^*(x) = \max_a Q^*(x, a)$. Thus, the value of $\hat{Q}^*$ is updated according to the following equation:

$$\hat{Q}^*(x, a) = (1 - \alpha)\hat{Q}^*(x, a) + \alpha\left(r + \gamma\hat{V}^*(y)\right) \tag{10.4}$$

Among them, $\alpha \in (0, 1]$ is a parameter that controls the learning rate, indicating how much trust is to be given to the corresponding update part.

The Q-learning algorithm uses TD(0) as an estimated factor for the expected return value. Note that the current estimate of the Q^* function is defined by a greedy strategy by $\pi(x) = \arg\max_a \hat{Q}^*(x, a)$. That is, the greedy strategy selects the action according to the maximum estimated Q value.

However, the first-order Q-learning algorithm does not clearly point out what action the agent should perform when each state updates its estimated value. In fact, all actions are likely to be executed by the agent. This means that in maintaining the best current estimate of the state, Q-learning allows for an arbitrary experiment. The action of following that statement is not important since the function is updated according to the optimal choice of the state surface. From this point of view, the Q-learning algorithm is not experimentally sensitive.

To eventually determine the best Q function, the agent must test all the available actions of each state many times. Experiments show that if the order of 10.8 is applied to all the Q of the state–action pair repeatedly, so that the update times of every Q of the state–action pair value will reach infinity, then Q^* will converge to Q^*, $\hat{V}^*$ will converge to V^*, and, as long as α decreases at a suitable rate to 0, the probability of convergence is 1. The backtracking of Q-learning is shown in Figure 10.8.

A typical single-step Q-learning algorithm is shown in Figure 10.9.

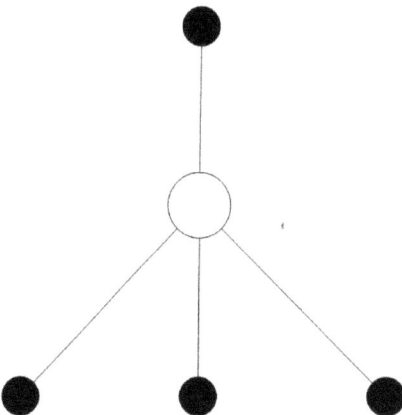

Figure 10.8: Backtracking of Q-learning.

Initialize Q(s,a) arbitrarily
Repeat(for each episode)
    Initialize s
    Repeat(for each step of episode):
        Choose a from s using policy derived from Q(e.g., ε-greedy)
        Take action a;observe r,s'
        $Q(s,a) \leftarrow Q(s,a) \leftarrow a[r - y \max Q(s',a') - Q(s,a)]$
        $S \leftarrow S'$
        until s is terminal

Figure 10.9: Single step of the Q-learning algorithm.

10.2.8 Improved Q-learning

The goal of Q-learning is to learn how to choose better actions or optimal actions according to external evaluation signals in a dynamic environment, which is essentially a dynamic decision learning process. When the agent does not understand the knowledge of the environment a little, it must learn by repeated experiments, and the efficiency of the algorithm is not high. Sometimes learning in an unknown environment will also involve a certain risk. One way to reduce this risk is to use the environment model. The environment model can be built from the experience gained in the previous execution of the related tasks. Using the environment model, it is easy to choose the action without the risk of being hurt.

The environment model is a function of the state and action (s_t, a) to the next state and the strengthening value (s_{t+1}, r). There are two ways to establish the model. First, in the initial stage of learning, the agent uses the data provided to build models offline. Second, the agent establishes or consummates the environmental model online in the process of interacting with the environment. The Q-learning algorithm based on experiential knowledge is a function E: $S \times A \rightarrow R$ with experience knowledge in the standard Q-learning algorithm. This function affects the selection of the agent action in the learning process, thus accelerating the convergence speed of the algorithm. Experience is represented by a four-tuple $\{s_t, a_t, s_{t+1}, r_t\}$, which indicates that an action a_t is executed at the state s_t, producing a new state s_{t+1} and getting an enhanced signal r_t.

The experiential function $E(s, a)$ in the improved algorithm records the relevant experience information about the execution action a under the state s. The most important problem of adding the experience function in the algorithm is obtaining the experience knowledge at the initial stage of learning, that is, defining the experience function $E(s, a)$. This depends mainly on the specific areas of the application of the algorithm. For example, in the agent path optimization environment, when the agent collides with the wall, the corresponding experience knowledge can be obtained. The agent obtains the experiential knowledge about the environment model online in the process of interacting with the environment.

The Q-learning algorithm based on experiential knowledge applies the empirical function to the agent action selection rules, and the action selection rules are as follows:

$$\pi(s_t) = \arg\max_{a_t}\left[\hat{Q}(s_t, a_t) + \varepsilon E_t(s_t, a_t)\right]$$

Here, ε is constant that represents the weight of the empirical function.

The Q-learning algorithm based on empirical knowledge is shown in Figure 10.10. Compared with the standard Q-learning algorithm, it is found that the algorithm is different from the strategy of action selection.

Initialize Q(s,a)
Repeat
    Visit the s state
    Select an action a using the action choice rule
    $\pi(s_i) = \arg\max[\hat{Q}(s_i,a_i) + \xi E_i(s_i,a_i)]$
    Receive r(s,a) and obesreve the next state s'
    Update the values of Q(s,a) according to:
    $Q(s,a) \leftarrow Q(s,a) + \alpha[r + y\max Q_i(s',a') - Q(s,a)]$
    Update the s to s' state.
Until some stop criteria is reached.
Where: $s \equiv s_t$, $s' \equiv s_{t+1}$, $a \equiv a_t$, $d \equiv a_{t+1}$

Figure 10.10: Q-learning algorithm based on the empirical knowledge.

Some of the primary ways of strengthening learning, of course, have been improved, but the mainstream is still the same. We have just introduced the basic principles, structure, and characteristics of reinforcement learning, as well as the MDP model that most classical reinforcement learning algorithms depend on. Then, we introduce the main elements of the reinforcement learning system: agent, environment model, strategy, reward function, and value function.

When the current state of the environment is down to a state, the probability and reward value of the state transfer depend only on the current state and selected actions, while the environment has the Markov property when it is close to the historical state and the historical action element. The reinforcement learning task that satisfies the Markov attribute is the decision-making process of Markov.

The main algorithms for reinforcement learning are DP, MC, TD, Q-learning, Q(λ)-learning, and Sarsa. If the agent does not need to learn the knowledge of the Markov decision model (R function and T function) during the learning process and learning the optimal strategy directly, this kind of method is called the model-independent method. In the course of learning, the method of learning model knowledge first and then deriving the optimization strategy according to the model knowledge is called the model-based method. DP and Sarsa are models based on common reinforcement learning algorithms. MC, TD, Q-learning, and Q(λ)-learning are typical model-independent methods.

In recent years, there have been more reinforcement learning about multi-machine learning. Therefore, although reinforcement learning was developed earlier, it still has huge prospects for further improvement.

10.3 Code implementation

Based on Sutton's book, "Reinforcement Learning," an example is found to be implemented as follows by Matlab.

1. Problem description
The Blackjack problem is also called the 21-point game. The rule of the 21-point game is introduced in brief as follows.

Twenty-one-point games usually use approximately one to eight side-cards. The player and the dealer receive two cards each. The player's cards are face up. The dealer has two cards: one card faces up (called the right numbers) and two cards face down (called dark cards). The number of poker points in your hands is calculated as follows: K, Q, J, and ten score 10 points. The A card can be counted as 1 or 11, which is decided by the player himself. All the remaining approximately two to nine cards are calculated according to their original face value. First, the player starts the card, and if the first two cards of the player are A and a 10-point card, the player is considered to have Blackjack. At this point, if the dealer does not have Blackjack, the game player will win two times sweepstakes (1/2). If the dealer has A brand, the game player can consider whether to buy insurance, the amount being half of the gambling chips. If the dealer has Blackjack, then the player takes back the insurance and wins directly. If the dealer does not have Blackjack, the player loses the insurance and the game continues. The players without Blackjack can continue to take the cards, which can be taken at random. The objective is to rely on 21 points as much as possible: the closer the better, with the best being 21. In the course of the card, if all the cards add up to more than 21 points, the player loses (Bust) and the game ends. If the player does not explode and decides not to sign again, the dealer opens his dark card. Generally at 17 or more than 17 points, one should no longer take cards; however, it is also possible at 15 or 16 points or even at 12 or 13 points to no longer take cards or even at 18 or 19 points to continue to take cards. If the banker explodes, he would lose. If he does not bust, he would have a big winning margin over a number of points. The same number of points is flat, and the bets can be brought back.

2. Program implementation
The program is implemented according to the 21-point rules, using reinforcement learning; the Matlab program is given as follows. The authors of the code are highly appreciated.

```
% BlackJack using Monte Carlo Policy
Current_sum=zeros(1,100)+12;
Dealer_show=0;
action=1;%0=stick 1=hit
Reward=0;
sum=10;
card=10;
i=1;
j=1;
Value_eval=zeros(sum,card)
Value_num=zeros(sum,card);
Valueval=0;
time=0;
for i=1:500000
%go on action if the action flag=1
while action==1
    time=time+1;
    j=j+1;
    %go out of the dealtplayer
    dealtplayer=randsrc(1,1,1:13)
    if dealtplayer>=10
        dealtplayer=10;
    end
    % do because of the ace and judge the Current_sum
    if (dealtplayer==1) && ((11+Current_sum(j))>21)
        Current_sum(j+1)=Current_sum(j)+dealtplayer;
    else if (dealtplayer==1)&&(11+Current_sum(j)<=21)
            Current_sum(j+1)=Current_sum(j)+11;
        else
            Current_sum(j+1)=Current_sum(j)+dealtplayer;
        end
    end
    if Current_sum(j+1)==20
        action=0;
    else
        if Current_sum(j+1)==21
            action=0;
            Reward=1;
        else if Current_sum(j+1)>21
                action=0;
                Reward=-1;
                Current_sum(j+1)=12;
```

```
            else
                aciton=1;
            end
        end
    end
end
    % do for the dealter
    dealtshow1=randsrc(1,1,1:13);
    if dealtshow1>=10
    dealtshow1=10;
    end
    dealtshow2=randsrc(1,1,1:13)
    if dealtshow2>=10
    dealtshow2=10;
    end
    if Reward~=-1
    if (dealtshow1==1) || (dealtshow2==1)
        dealtshow2=11;
    end
    dealershow=dealtshow2+dealtshow1;
    if dealershow==Current_sum
        Reward=0;
    else if dealershow>Current_sum
            Reward=-1;
        else
            Reward=1;
        end
        end
        end
        % ti sum of the Value
        for j=1:100
        Value_eval(Current_sum(j)-11,dealtshow1)=Value_eval
        (Current_sum(j)-11,
            dealrshow1)+Reward;
        Value_num(Current_sum(j)-11,dealtshow1)=Value_num
        (Current_sum(j)-11,
            dealershow1)+1
end
Reward=0;
action=1;
j=1;
```

```
Current_sum=zeros(1,100)+12;
end
%eveage of the sum
Value_eval=Value_eval./Value_num
```

Bibliography

Albert, J. (2009). *Bayesian Computation with R, Second edition*. New York, Dordrecht, Springer.

Bishop, C. M. (2006). *Pattern recognition and machine learning*. New York, Springer.

Busoniu L, B. R. (2017). *Reinforcement learning and dynamic programming using function approximators*. Boca Raton, London, CRC Press.

Chatzis, S. P. (Dec. 2010). Hidden Markov Models with Nonelliptically Contoured State Densities. *IEEE Transactions on Pattern Analysis and Machine Intelligence*, S. 32 (12): 2297–2304.

Choudhuri N, G. S. (2005). Bayesian methods for function estimation. *Handbook of Statistics*, 25: 373–414.

Chunyu Xie, C. L. (2018). Memory attention networks for skeleton-based action recognition. *Proceedings of International Joint Conference on Artificial Intelligence*. IEEE.

Cortes C, V. V. (1995). Support-vector networks 1995. *Machine learning*, 20(3): 273–297.

Donoho, D. (2006). Compressed sensing. *IEEE Transactions on Information Theory*, S. 52 (4): 1289–1306.

Drucker H, B. C. (1997). Support vector regression machines. *Advances in neural information processing systems*, S. 9: 155–161.

Duda R O, H. P. (2012). *Pattern classification*. New York: John Wiley & Sons.

Elgammal, A. (2005). *CS 534 Spring 2005: Rutgers University Object Detection and Recognition*. Von https://www.cs.rutgers.edu/~elgammal/classes/cs534/lectures/ObjectRecognition.pdf abgerufen

Gu J, L. C. (2019). Projection Convolutional Neural Networks for 1-bit CNNs via Discrete Back Propagation. *American Association for Artificial Intelligence*, S. 33: 8344–8351.

Haussler D, W. M. (1993). The probably approximately correct (PAC) and other learning models. *Foundations of Knowledge Acquisition, Springer*, S. 195: 291–312.

Haussler, D. (1990). *Probably approximately correct learning*. University of California, Santa Cruz, Computer Research Laboratory.

Hsu C W, L. C. (2002). A comparison of methods for multiclass support vector machines. *IEEE transactions on Neural Networks*, S. 13(2): 415–425.

Huang, D. (1996). *The Theory of Neural Network Pattern Recognition System*. Beijing: Electronic Industry Press.

I.T., J. (2002). *Principal Component Analysis, Series: Springer Series in Statistics, 2nd ed.* New York: Springer.

K., H. T. (1995). Random decision forests. *IEEE Proceedings of 3rd international conference on document analysis and recognition* (S. 278–282).

Kamiński B, J. M. (2018). A framework for sensitivity analysis of decision trees. *Central European journal of operations research*, 26(1): 135–159.

Kearns M J, V. U. (1994). *An Introduction to Computational Learning Theory*. Cambridge, Massachusetts: MIT Press.

LeCun Y, B. Y. (2015). Deep learning. *Nature, 2015*, S. 521 (7553): 436.

LI C, X. C. (2018). Deep Fisher discriminant learning for mobile hand gesture recognition. *Pattern Recognition*, 77: 276–288.

Li, H. (2012). *Statistical Learning Method*. Beijing: Tsinghua University Press.

Linlin Yang, C. L. (2017). Image reconstruction via manifold constrained convolutional sparse coding for image sets. *IEEE Journal of Selected Topics in Signal Processing*, S. 11(7): 1072–1081.

Luan S, C. C. (2018). Gabor convolutional networks. *IEEE Transactions on Image Processing*, S. 27(9): 4357–4366.

https://doi.org/10.1515/9783110595567-012

Luo, S. (2006). *Information Processing Theory of Visual Perception System*. Beijing: Electronic Industry Press.

M., E. (2010). *Sparse and redundant representations: from theory to applications in signal and image processing*. New York, Dordrecht, Springer Science & Business Media.

M., M. T. (2003). *Machine Learning, the first edition*. Beijing: Mechanical Industry Press.

M., W. H. (1969). *Principles of operations research: with applications to managerial decisions*. Englewood Cliffs, NJ: Prentice-Hall.

Mitchell., T. M. (2013). *Machine Learning. Junhua Zen, Yinkui Zhang translate*. Beijing: Machinery Industry Press.

Natarajan, B. K. (1991). Machine Learning, A Theoretical Approach. *Journal of Japanese Society of Artificial Intelligence, Morgan Kaufmann Publishers*.

R., R. (2009). *AdaBoost and the super bowl of classifiers a tutorial introduction to adaptive boosting*. Berlin: Freie University. Von http://www.inf.fu-berlin.de/inst/ag-ki/adaboost4.pdf abgerufen

Richard S Sutton, A. G. (1998). *Reinforcement Learning: An Introduction*. Cambridge, MA: MIT Press.

Russell S J, N. P. (2016). *Artificial Intelligence: A Modern Approach*. Malaysia; Pearson Education Limited.

Sotirios P. Chatzis, D. K. (July 2012). Visual Workflow Recognition Using a Variational Bayesian Treatment of Multistream Fused Hidden Markov Models. *IEEE Transactions on Circuits and Systems for Video Technology*, S. 22 (7): 1076–1086.

Sutton R S, B. A. (2018). *Reinforcement learning: An introduction*. Cambridge, London, MIT Press.

V., V. N. (2004, 1998). *Statistical learning theory. Xu JH and Zhang XG. translation*. Beijing: Publishing House of Electronics Industry.

Valiant, L. (1984). A theory of the learnable. *Communications of the ACM*, 27(11): 1134–1142.

Valiant, L. (2013). *Probably Approximately Correct: NatureÕs Algorithms for Learning and Prospering in a Complex World*. New York, Basic Books (AZ).

Vasilescu M A O, T. D. (2003). Multilinear subspace analysis of image ensembles. *Computer Society Conference on Computer Vision and Pattern Recognition* (S. 2: II–93.). IEEE.

Wang L, S. M. (2008). On the margin explanation of boosting algorithms. *Conference on Learning Theory* (S. 479–490).

Wang X, L. C. (2019). Taylor convolutional networks for image classification. *IEEE Winter Conference on Applications of Computer Vision* (S. 1271–1279). IEEE.

Wang X, Z. B. (2018). Modulated convolutional networks. *Proceedings of the IEEE Conference on Computer Vision and Pattern Recognition* (S. 840–848). IEEE.

Wikipedia. (2019). Von Wikipedia: http://www.wikipedia.org/ abgerufen

Xie C, L. C. (2019). Hierarchical residual stochastic networks for time series recognition. *Information Sciences*, 471: 52–63.

Zhang B, P. A. (2018). Manifold constraint transfer for visual structure-driven optimization. *Pattern Recognition*, S. 77: 87–98.

Zhang, X. (2010). *Pattern Recognition*. Beijing: Tsinghua University Press.

ZhaoQi, B. (2012). *Pattern Recognition*. Beijing: Tsinghua University Press.

Zheng, N. (1998). *Computer Vision and Pattern Recognition*. Beijing: National Defense Industry Press.

Zhou, Z. (2016). *Machine Learning*. Beijing: Tsinghua University Press.

Index

AdaBoost 1, 75, 77, 78, 79, 80, 82, 84, 85, 87
AutoEncoder 109, 116, 117, 118, 119

Backward Propagation 109, 112
Bagging 75
Baum-Welch algorithm 59, 60
Bayesian classification 54
Bayesian decision 16, 47, 48, 49
Bayes' theorem 47
Boosting 75, 122, 123

Cascade 82, 84
Compressed sensing 89, 90, 94, 97
conditional probability density 16, 47, 48, 49, 50
connectionist learning 6, 10, 127

decision tree 1, 2, 10, 11, 33, 34, 35, 36, 37, 38,
    39, 40, 42, 44, 45, 54, 75, 86, 88
Deep learning 109, 125
Dynamic programming 130

empirical risk 17, 18, 19, 20, 21, 22, 23
expected risk 15, 16, 17, 18, 19, 21

Face Recognition 95, 107
finite space 25
Forward-backward algorithm 59, 60

global optimization 42

HMM 57, 58, 59, 60
HOG 116

ID3 algorithm 33, 40, 42
infinite space 25

kernel function 66, 67, 68, 69, 74

Lagrange coefficient 65
linear classifier 18, 23, 30, 63, 69, 76, 80
log-likelihood function 50

Machine learning 3, 4, 5, 6, 13, 15, 87
Markov chain 56, 57, 58, 59
Markov process 56, 59, 128

Monte Carlo 130, 131, 132, 137

Naive Bayesian 51, 52, 54
Nyquist–Shannon sampling theorem 89

Object Detection 75, 80, 85
Objective function 36

PAC 1, 25, 26, 27, 28, 29, 31
pattern classification 47
principal component analysis 99, 100, 102,
    103, 105, 106
probably approximately correct learning 25

Q-Learning 132, 133, 134, 135

Reinforcement learning 1, 6, 12, 13, 127, 128,
    130, 131, 132, 136, 137
Restrict Boltzmann Machine 116
RIP criterion 93

Shannon information entropy 37
SIFT 116
Signal Reconstruction 93
Sparse Representation 91
statistical inference 47
Statistical pattern recognition 14
structural risk minimization 23, 24, 63, 74
Subspace Learning 99
supervised learning 63, 110, 116, 117, 118, 127
Support Vector Machine 1, 2, 14, 53, 63, 66,
    67, 68, 69, 70, 71, 73, 74, 96, 116

Temporal Difference 131
training data 25, 27, 51, 63, 67, 116, 120, 123
transition probability matrix 57

unsupervised learning 116, 127

value function 127, 129, 131, 132, 136
variance 14, 49, 50, 51, 99, 100, 103, 104, 105,
    106, 107, 120, 121
VC dimension 19, 20, 21, 22, 23, 29, 30, 63
VC trust 21
Viterbi algorithm 59, 60

https://doi.org/10.1515/9783110595567-013

www.ingramcontent.com/pod-product-compliance
Lightning Source LLC
Chambersburg PA
CBHW082034230326

41598CB00081B/6507